Survival in space

Survival in space

Medical problems of manned spaceflight

Richard Harding

ROUTLEDGE

London and New York

First published 1989
by Routledge
11 New Fetter Lane, London EC4P 4EE
29 West 35th Street, New York, NY 10001

Typeset by Witwell Ltd, Southport
Printed and bound in Great Britain by
Biddles Ltd, Guildford and King's Lynn

British Library Cataloguing in Publication Data

Harding, Richard M.
    Survival in space: medical problems of
    manned spaceflight.
    1. Astronauts. Physiology. Effects of
    space flight
    I. Title
    612′0145

Library of Congress Cataloging in Publication Data

Harding, Richard M.
    Survival in space.
    Bibliography: p.
    Includes indexes.
    1. Space medicine. I. Title. (DNLM: 1. Space
Flight. WD 750 H261s)
RC1135.H37 1989 616.9′80214   88–32091

ISBN 0–415–00253–2

For my mother and father

Contents

Contents

Illustrations

Plates

Preface

Space is the ultimate hostile environment. It is the purpose of this book to help explain how the physiologists, physicians, and engineers involved in the manned spaceflight programmes have sought to provide the necessary protection for man to survive in space.

The book is divided into three parts, although each clearly interrelates with the others. Part 1 consists of a brief discussion of the nature of space, followed by a short historical review of the manned spaceflight programmes. Part 2, 'Putting Man into Space', addresses the *immediate* hazards facing any space traveller, and the methods adopted to provide the necessary protection. Although the distinction from Part 2 is clearly not well defined, I have chosen to call Part 3 'Keeping Man in Space', and to consider therein the consequences of *prolonged* space travel upon the human body and mind. Thus, the chapters in this final part deal with the physiological consequences of exposure to microgravity (weightlessness) and their implications; and with the increasingly important psychological aspects of the long duration space missions planned for the future. The involvement of women in space is also considered. A third chapter discusses the likely changes and improvements needed, in the fields covered by earlier chapters, to ensure our continuing survival in space, and for however long we wish to journey there.

Despite the tragic loss of the Shuttle Challenger, the manned exploration of space is today enjoying a renascence and the need to survive in space is perhaps even more relevant as ever increasing numbers of people journey away from the Earth for long periods. Large, continuously manned space stations are at an advanced stage of planning by both the Russians and the Americans, and Shuttle-type vehicles are intended to fly almost weekly. And, at long last, it appeared that Great Britain was becoming more actively involved: November 1985 saw the long-overdue establishment of a British National Space Centre. The first fruit of its labours appeared in March 1987 as the proposed United Kingdom 'Space Plan', and included a contribution to Columbus, the European Space Agency's own space station project. It is most unfortunate that funding for the 'Plan' has so far been rejected by the government. In addition, of course, 1986 was destined to be the year in which the first Briton travelled into space: let us hope that, despite the apparent eclipse of the United Kingdom's role in space, the event will not be long delayed, and that it will be the first of many.

RMH
Farnborough

Postscript: On 29 September 1988, the United States re-established its manned spaceflight programme, after a 32-month hiatus, with the successful launch of the Shuttle Discovery. The five man crew returned to Earth safely four days later. The day of the launch was chosen by the Russians to unveil to the West their version of a reusable manned spacecraft: externally a virtual twin to the American Shuttle. On the same day, President Reagan signed an accord to build a manned space station by the late 1990s with eleven other nations. Manned spaceflight rightly remains an active and serious endeavour.

Foreword

Books about the manned space program tend to be one of two kinds. The first kind are written by actors in the drama, and are heavy on narrative and color but light on specific facts. The second kind are compendia of scientific and technical data, usually by many authors, useful for reference but not very readable.

Dr Harding's book, *Survival in Space*, has managed to bridge this gap. He presents a detailed narrative of the adventure while, at the same time, relating in detail the approach and solution to the many and complex medical and life support problems attending the invasion by man of this most abnormal environment.

The book is filled with facts, some of which I did not know or have forgotten, and will serve as an extremely useful reference volume to both professionals and laymen. But it also tells a story – or, at least, the beginning of a story. We all hope that many chapters remain to be written.

Joseph P. Kerwin,
Houston, Texas,
March 1989.

Acknowledgements

It is a pleasure to thank my friends and colleagues, at Farnborough and elsewhere, for their help with this project. I am especially grateful to Air Vice Marshal Peter Howard, lately the Dean of Air Force Medicine and Commandant of the Royal Air Force Institute of Aviation Medicine, for his careful scrutiny of the manuscript and invaluable criticism. I would also like to thank the publishers and particularly my editor, David Stonestreet, for inviting me to write this book and for help and encouragement throughout. I am grateful to those officials at NASA who knowingly, or unknowingly, have made the book possible – for it must be apparent to all that NASA has somewhat of a monopoly in this field, and I have drawn extensively on its written and pictorial publications. With regard to the latter, I am especially indebted to Lisa Vasquez. Finally, for her patience, interest, and support, yet again I thank my wife, Letitia.

The photographs appear by courtesy of the National Aeronautics and Space Administration.

Abbreviations

ACS	Atmospheric Control System
ADH	Antidiuretic Hormone
ALSEP	Apollo Lunar Surface Experiment Package
ALSV	Air-Launched Sortie Vehicle
AMU	Astronaut Manoeuvring Unit
ASTP	Apollo-Soyuz Test Project
BIS	Bioinstrumentation System
CELSS	Closed Ecological Life Support System
CM	Command Module
CNS	Central Nervous System
CSM	Command and Service Module
CSR	Critical Supersaturation Ratio
CWG	Constant Wear Garment
ECF	Extracellular Fluid
ECG	Electrocardiogram
ECS	Environmental Control System
EEG	Electroencephalogram
ELSS	Extravehicular Life Support System
EMG	Electromyogram
EMU	Extravehicular Mobility Unit
ENT	Ear, Nose, and Throat
EOG	Electro-oculogram
ESA	European Space Agency
EVA	Extravehicular Activity
FCS	Faecal Containment System
FSS	Flight Support Station
GEO	Geostationary Earth Orbit

HHMU	Hand Held Manoeuvring Unit
HOTOL	Horizontal Take-off and Landing
HSP	Health Stabilization Programme
ICF	Intracellular Fluid
IR	Ionizing Radiation
ITMG	Integrated Thermal Micrometeoroid Garment
IVA	Intravehicular Activity
LBNP	Lower Body Negative Pressure
LCG	Liquid Cooling Garment
LEO	Low Earth Orbit
LET	Linear Energy Transfer
LEVA	Lunar Extravehicular Visor Assembly
LM	Lunar Module
LQP	Lunar Quarantine Programme
LRL	Lunar Receiving Laboratory
LRV	Lunar Roving Vehicle
LSS	Life Support System
MASTIF	Multiple Axis Space Test Inertia Facility
MISS	Man in Space Soonest
MMU	Manned Manoeuvring Unit
MQF	Mobile Quarantine Facility
MSC	Manned Spaceflight Centre
NASA	National Aeronautics and Space Administration
OTF	Orbital Test Flight
OPS	Oxygen Purge System
OWS	Orbital Workshop
PG	Pressure Gloves
PGA	Pressure Garment Assembly
PHA	Pressure Helmet Assembly
PLSS	Portable Life Support System
PNS	Peripheral Nervous System
POS	Portable Oxygen System
PRS	Personal Rescue Sphere
PTC	Passive Thermal Control
RBE	Relative Biological Effectiveness
REM	Rapid Eye Movements

SI	Système Internationale
SM	Service Module
SOMS	Shuttle Orbiter Medical System
SRB	Solid Rocket Booster
SSA	Spacesuit Assembly
STS	Space Transportation System
SWVP	Saturated Water Vapour Pressure
TLSA	Torso-Limb Suit Assembly
UCTA	Urine Collection and Transfer Assembly
URA	Urine Receptacle Assembly
UTS	Urine Transfer System
VCM	Ventilation Control Module
WMS	Waste Management System

A note about units

A brief explanation of some of the units of measurement used in this book may be of help at this stage.

Measurement of pressure may be expressed in a wide variety of units, such as millimetres or centimetres of mercury, as torr, as pounds per square inch, as bars, and as millimetres, centimetres, or inches of water. The Système Internationale (SI) unit is the Pascal (Pa)! Doctors seldom readily accept or adapt to change without good reason, and the traditional unit for the expression of pressure in human physiology remains that of the easily understood millimetre of mercury (mmHg): I have perpetuated this tradition, although the SI equivalent in kilopascals (kPa) is also given.

Height is the term used to denote the distance of an aircraft or spacecraft above ground level, while altitude is the term used to denote the distance above mean sea level. In the world of aerospace, such distances are mandatorily expressed in feet. This convention is obeyed in this book but the SI conversion to metres or kilometres is also given.

For other units of measurement, I have tried in most cases to use the SI, with conversion to the more familiar and traditional units where these may aid comprehension.

Part one

Introduction

Since Wednesday 12 April 1961, when Cosmonaut Yuri Gagarin spent 108 minutes in space orbiting the Earth in Vostok 1, more than 200 men and women of twenty nations have spent a total of more than 7,700 man-days in space. This remarkable achievement in just over one-quarter of a century represents one of civilization's greatest triumphs. Although the cosmonauts and astronauts are, quite correctly, the heroes and heroines of this adventure, many thousands of scientists from many disciplines have played a vital role in the success of the manned spaceflight programmes. This introduction provides a brief description of the physical nature of 'space' as currently understood, followed by a short historical survey of the Russian and American manned spaceflight programmes, with particular emphasis on the medical and survival aspects.

Chapter one

The nature of space

According to the 'Big Bang' theory, the Universe began as a massive serendipitous explosion within a large blob of matter, the so-called primeval atom or cosmic egg, some 18,000 million years ago. Our galaxy, including the Solar System, was formed from coalescing material 1,000 million years later. Life on Earth probably began about 3.3 million years ago and it really is inconceivable that, among all the billions of stars within the universe, ours is the only one to support life. It is this likelihood, coupled with humanity's insatiable curiosity, that will continue to impel us to explore the secrets of the universe. Our forebears have dreamt of so doing for centuries but today we are equipped with the expertise and technology to travel beyond Earth's immediate vicinity (our biosphere) and to start doing as well as dreaming.

The transition to space

The biosphere is that part of the universe in which life can be sustained without artificial support: at present, clearly, this comprises the Earth's land and sea masses (the lithosphere and hydrosphere respectively) and the mass of air (the atmosphere) above them. The atmosphere is composed of a life-giving mixture of gases held in place around the globe under the influence of gravity. This influence declines as the distance away from the Earth's surface increases. At the same time, solar heating becomes more and more effective: the heat of the Sun expands the gases in the atmosphere, so driving their molecules further and further apart and causing the density and pressure of the air to decline with altitude. Such a reduction means that the native lowlander is unable to survive for long periods above 10,000 feet (3,048 m), while even his native highlander cousins can only adapt to sustained life for a further 8,000 feet (2,438 m). At an altitude of 40,000 feet (12,192 m) no human can survive without an atmosphere composed entirely of oxygen, and at an altitude of 60,000 feet (18,288 m) a spacesuit or pressure cabin is an absolute requirement for survival. For the purposes of this book, therefore, this limitation is perhaps the best definition of the beginning of what could be called Physiological Space. There is, however, even at 60,000 feet, a finite

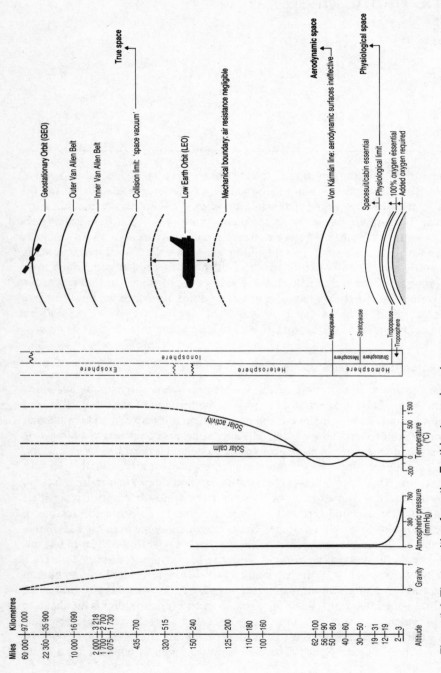

Figure 1.1 The transition from the Earth's atmosphere to space

atmospheric pressure (of 54 mmHg (7.2 kPa)) and gravity is still exerting a force of well over 99 per cent of that at sea level.

Figure 1.1 is a graphical representation of the physical transition from the Earth's surface to space.

By 262,000 feet (50 miles, 80 km) atmospheric pressure has fallen to <0.25 mmHg (0.03 kPa), and collisions between molecules (that is, the physical nature of pressure) have become so uncommon that normal aerodynamic forces cease to be effective. Above this point, which may be regarded as the beginning of Aerodynamic Space and which is termed the von Kármán line, rockets or reaction jets, powered solely by internally carried fuel, become the only means available for manoeuvrability.

It is not until an altitude of 435 miles (700 km) is reached, where collisions between molecules become such rare events as to be almost entirely unknown and a total vacuum exists, that True Space can be said to begin. Earth's gravity, however, still exerts a considerable attraction, even at this altitude, and it is not until altitude of 1,700 miles (2,735 km) that it is halved, and an altitude of 60,000 miles (96,540 km) that it is one-fifth of that at sea level. Theoretically, no spacecraft launched from Earth would be entirely free of its gravitational influence until it had travelled several million miles away. Long before then, of course, other bodies, such as the Moon, would begin to exert a greater gravitational effect as the craft approached.

The atmosphere, as well as supporting life on Earth directly, by virtue of its oxygen and carbon dioxide content, also provides a vital shield against the unwelcome intrusion of all forms of radiated energy from space. The most obvious of these is radiant heat from the Sun, which modifies and is itself modified by the various layers of the atmosphere. The lowest layer, the troposphere, extends from the Earth's surface upwards for about 11 miles (17.7 km) at the equator and for about 5 miles (8 km) at the poles. The carbon dioxide and water vapour within this layer are unable to absorb solar heat because its wavelength is too short. Instead, the Earth itself absorbs the energy and re-radiates it back into the troposphere to provide an acceptable thermal environment over most of our planet. As the heat rises and is dissipated, a declining temperature gradient is seen in the troposphere, from an average 15°C at sea level to –83°C at its upper limit (tropopause) over the equator and to –53°C over the poles. It is this gradient which, in part, is responsible for the weather.

The next layer is called the stratosphere and is characterized by an increasing temperature gradient which is gradual at first but then intensifies so that the temperature at its upper limit, at about 30 miles (48 km) and called the stratopause, has risen to –3°C or even higher. This increase in temperature is a consequence of the absorption of ultraviolet light from the Sun by ozone within the stratosphere. Ozone (O_3) is the triatomic form of oxygen and a strong oxidizing agent. Within this so-called ozonosphere, ultraviolet light breaks down normal molecular (diatomic) oxygen (O_2) into its two constituent atoms, which are then free to combine with residual molecules of oxygen to form ozone. In absorbing further ultraviolet energy, ozone is itself broken down and the cycle is repeated, so serving to protect the atmosphere below from the harmful effects of ultraviolet radiation.

The third layer, the mesosphere, exhibits a rapid decline in temperature to about $-113°C$ at an altitude of 55 miles (88.5 km). Thereafter, a number of imprecise layers exist, collectively comprising the heterosphere (a transition zone) and then the exosphere, throughout which there is a progressive increase in temperature to a level which is dependent upon the activity of the Sun. Temperatures can exceed $1,500°C$ during days of maximum solar activity but may be as low as $227°C$ during nights of solar calm. Thus, the lower layers of the atmosphere may be regarded as a thermally protective layer around the globe, without which the effects of the Sun would be catastrophic: the Earth's surface would boil. An obvious extension of this is that temperature control within spacecraft is critical if the potential extremes of unprotected exposure are to be avoided; and this problem will be addressed in Chapter 6. In the environment of space, the concept of temperature is a *non sequitur*: since there are no molecules to collide, heat cannot be produced or attenuated. The cabin of a spacecraft, or interior of a spacesuit, is therefore exposed directly to transmitted heat or cold depending on its orientation to the Sun.

Similarly, the effects of ionizing radiation, whether from within our own Solar System or from beyond it, may be devastating. X-rays, gamma rays, and atomic particles of many types (including electrons, protons, and neutrons) all bombard the Earth continuously but are largely dispersed by the atmosphere. Again, spacecraft enjoy no such luxury and need to be positively protected, and this aspect will be discussed further in Chapter 5. Ionizing radiation is present throughout the heterosphere and exosphere, and indeed it is the electrical properties of this radiation which best define the physical nature of the space environment in these regions.

Thus, space does not begin abruptly at some arbitrary point above the Earth's surface. The atmosphere becomes progressively thinner and more hostile as distance from Earth increases but is still measurable, in terms of density and hence pressure, at an altitude of 23 miles (37 km). Of course, for the vast majority of people, the existence of an atmosphere at such altitudes is merely academic since it is open to relatively small numbers of the Earth's population to travel at more than even a few feet above its surface; and to an extremely elite minority, so far, to voyage beyond 60,000 feet (18.3 km). In other words, for most people on Earth, space might just as well begin at 10 feet (3 m) above its surface!

The problems of gravity

The same gravitational force which retains the atmosphere about the Earth conspires to prevent objects such as spacecraft both from leaving to explore beyond and from returning safely. Chapter 4 will expand upon the problems of overcoming gravity during both launch and re-entry of spacecraft but some brief explanation of celestial mechanics is appropriate here.

A rocket destined to leave the atmosphere must generate an upward thrust which is at least greater than the downward attraction of gravity. That is, an acceleration greater than 32 feet. $\sec^{-2}$ (9.8 m. $\sec^{-2}$), the acceleration due to gravity, must be achieved. For example, a rocket weighing n pounds which develops a thrust at take-off of $2n$ will accelerate away from Earth at 32 feet. $\sec^{-2}$. The greater the thrust, the more quickly will the rocket achieve its purpose, but the loads on the rocket structure and that of its payload, including man, will be correspondingly and perhaps unacceptably elevated.

As the rocket gets higher and burns off fuel, it gets lighter and will accelerate further. This is the rationale behind the staging of rocket launch vehicles (a concept described, separately, by all three 'fathers of rocketry': Tsiolkovsky in Russia, Goddard in America, and Oberth in Germany) whereby progressive jettisoning of empty rocket stages, followed by ignition of the next stage, results in a reduction in weight and an increase in acceleration.

A further energy saving can be made by taking advantage of the Earth's own rotational movement to reduce the velocity required to attain orbit. At the equator this rotational surface velocity amounts to 1,500 feet. $\sec^{-1}$ (450 m. $\sec^{-1}$) in an easterly direction, and so this is the bearing on which most spacecraft are ultimately sent. Launching to the north or south loses this benefit. It does, however, achieve a polar orbit which has distinct advantages of its own: in polar orbit, a spacecraft or satellite will cross all points of the globe once every few revolutions, so creating unique opportunities for observation.

In order to achieve any orbit about the Earth, a rocket must be launched vertically and then gradually tilted (usually easterly) so that it attains a flight path parallel to the Earth's surface at the moment when orbital velocity at the required altitude is reached. If the rocket engine is stopped at precisely this time, and provided that the spacecraft is beyond the influence of atmospheric drag (that is, above the von Kármán line), it will orbit the Earth almost indefinitely as the centrifugal force of its motion (tending it away from the centre) exactly balances the centripetal force of gravity attracting it back to Earth. The velocity of a spacecraft in Earth orbit will vary with the altitude at which the craft is intended to operate. Thus for an orbit at 125 miles (200 km) a velocity of 17,900 miles. h^{-1} (28,800 km. h^{-1}) is required. At this altitude, the period (time) for one orbit of the Earth will be two hours. Earth-orbiting manned spacecraft, such as the Russian Salyut space stations and the American Shuttles, operate at or just above this altitude and are said to be in Low Earth Orbit (LEO). For higher orbits, the velocity required is less and the orbital period correspondingly greater. At an altitude of 22,300 miles (35,880 km), the orbital velocity is just 7,000 miles. h^{-1} (11,263 km.h^{-1}) and the period is twenty-four hours. Since this is also the time it takes for the Earth to rotate once, spacecraft in this type of orbit are held apparently motionless over a set point on the Earth's surface. This Geostationary Earth Orbit (GEO), as it is known, is particularly useful for communications satellites.

Finally, a launch velocity of some 25,950 miles. h^{-1} (41,753 km. h^{-1}) is required in order to achieve escape from the influence of the Earth's gravitational pull and so to travel to the Moon or other planets. The timing of such a launch is critical since it involves extremely complicated mathematical considerations of the relative motion of the three bodies involved: the Earth, the spacecraft, and the Moon or target planet. The complexity of these calculations was evident during the only Earth Escape missions so far undertaken by manned spacecraft, the American Apollo flights to the Moon. For this programme, the Saturn V three-stage rocket first lifted the Apollo into Earth orbit at about 116 miles (187 km) and a velocity of 17,427 miles. h^{-1} (28,040 km. h^{-1}). At precisely the right moment, the rocket engines were fired again to propel the spacecraft on the correct trajectory for the Moon. This manoeuvre increased the velocity to 24,200 miles. h^{-1} (38,938 km. h^{-1}) initially, but the continuing influence of Earth's gravity had slowed the spacecraft to 2,040 miles. h^{-1} (3,282 km. h^{-1}) as it approached the Moon. The Moon's own gravitational field then took over and accelerated the Apollo again to a velocity of 5,600 miles. h^{-1} (9,010 km. h^{-1}) as it travelled behind the Moon. Retro-rockets were needed to reduce this velocity to that required for lunar orbit (3,680 miles. h^{-1} (5,921 km. h^{-1})). Similar computations were then required for the lunar landing and, of course, the whole cycle was repeated for the return journey.

Re-entry is that part of a space mission during which return to the Earth's atmosphere is accomplished. The atmosphere, however, acts as a two-edged sword since its increasing density, while being used as a braking or retarding influence, also causes immense frictional heating over the blunt leading surface of the spacecraft. This heat, which is caused by the rapid movement of nitrogen and oxygen molecules over the surface, is dissipated by deliberate vaporization of some components of the ablative heat shield. In melting and vaporizing, these special plastics carry away the so-called boundary layer heat, while further insulating materials behind the shield protect the pressure cabin itself. Even so, the heat shield of a returning Apollo was subjected to temperatures of nearly 3,000°C (5,000°F)! The temperature generated during this phase is acutely dependent upon the angle of re-entry, as is the acceleration profile. Thus, an angle of –6.2° ± 1.0° with respect to the Earth's horizon is the ideal and safe approach. If the angle is too shallow, the spacecraft will bounce off the outer layers of the atmosphere; while if it is too acute, the spacecraft may not survive the extreme temperatures even if the crew were able to tolerate the extreme accelerations. As it is, the narrow angle of entry allows an appropriate degree of deceleration while avoiding excessive heating.

To voyage beyond the bounds of Earth is thus a formidable and hazardous undertaking. That we have chosen to attempt it was inevitable and there is equally no doubt that we will for ever continue to challenge the cosmos:

The history of the human race is a continuous struggle from darkness toward light. It is, therefore, of no purpose to discuss the use of knowledge – man wants to know. And when he ceases to do so, he is no longer man.

(Fridtjof Nansen)

Chapter two

The manned spaceflight programmes

The Russian manned spaceflight programme

The year 1957 was International Geophysical Year and on 4 October the Russians launched Sputnik 1 into Earth orbit; it was a small aluminium sphere 58 cm (23 in) in diameter weighing nearly 84 kg (185 lb). The space age had truly begun. Just one month later a dog called Laika was launched into Earth orbit on board Sputnik 2, a cone-shaped satellite weighing 504 kg (1,111 lb). The correct functioning of the dog's life support system (LSS) was monitored from Earth by telemetry, as were the animal's vital signs – heart rate, blood pressure, and respiration – until she died several days later from lack of oxygen. Although several other 'physical sciences' Sputniks were launched subsequently, dogs, rats, mice, and flies were all to be sent aloft in a series of five 'biological' Sputnik launches before Gagarin undertook his historic flight. The animals in these later satellites were successfully recovered to Earth and were thus the first real space travellers.

Vostok

Then came Gagarin's single-orbit flight in Vostok 1. The Vostok spacecraft (Vostok means 'East' in Russian) consisted of an upper spherical capsule 2.3 m (7.5 ft) in diameter, containing the LSS, instrumentation, and an ejection seat, and a lower conical module containing support services such as batteries, gas bottles, and retro-rockets. A pressurized spacesuit was worn throughout the flight and at the end of the orbit, after re-entry, Gagarin ejected from his capsule at an altitude of about 22,000 feet (6,700 m) to complete the final stages of descent by parachute! Not surprisingly, the cosmonaut's well-being was of prime concern and, as with *all* subsequent spaceflights, a safe launch, with complete protection against the hazards of space while aloft, followed by a safe and accurate re-entry and descent was the only acceptable course of events. The effects of microgravity, however, were far from clear at this stage, although the animal experiments had indicated that normal bodily function could be maintained. Consequently, Gagarin was very heavily instrumented and medical monitoring included continuous measurement of heart rate, respiration, electrical activity of the brain (electroencephalogram – EEG), eye movements (electro-oculogram – EOG), muscle activity (electromyogram – EMG), and temperature responses. The data were transmitted back to Earth, i.e. telemetered, for observation and analysis. All measured functions remained normal throughout Gagarin's flight, as they did for the second Russian in space: Cosmonaut Gherman Titov.

Titov's flight in Vostok 2 lasted over one day and involved seventeen orbits of the Earth. He was the first man to sleep in space and he was also the first to experience space motion sickness. Although this problem had been predicted before any manned spaceflight had been undertaken, its importance and magnitude had yet to be appreciated.

Four more Vostok spacecraft were launched, in two pairs; the first in August 1962 and the second in June 1963. These paired missions were important stepping stones towards an ability to rendezvous and dock in space and, although the latter was never attempted in these flights, the spacecraft did approach to within four miles (6.5 km) of each other. Vostok 6 was also of major importance since its pilot was a woman, a 26-year-old ex-factory worker named Valentina Tereshkova. Her biomedical monitoring revealed no major differences from the responses to spaceflight seen in her male colleagues, although at times during her 70-hour flight she suffered severe nausea and disorientation and was extremely unhappy. In November 1963, Cosmonaut Tereshkova married Cosmonaut Andrian Nikolayev, the pilot of Vostok 3, and the following year gave birth to a healthy baby girl: the first true child of the space age. The six Vostok flights took the Russian experience in space to just over twenty-one man-days.

Voskhod

There followed two flights of the Voskhod spacecraft. These craft closely resembled the Vostoks but the ejection seat was removed so that the crew complement could be increased. With the option of ejecting from the capsule denied them, the occupants were obliged to accept a somewhat hard recovery on to land. As in the Vostok missions, virtually all spacecraft systems were automated, with the cosmonaut acting as an intelligent passenger and research animal (although he could fly the vehicle manually in an emergency). Voskhod 1 (October 1964) carried three cosmonauts, including a physician/physiologist, while Voskhod 2 (March 1965) was a two-man mission during which the first extravehicular activity (EVA) was undertaken. A mobile LSS was required for this, although the cosmonaut remained tethered to his spacecraft by an umbilical connection carrying the spacesuit supplies. In addition, the crews of the Voskhods dispensed with the requirement to wear spacesuits continuously and were able to carry out even more biological monitoring, including that of lung function, vestibular function, and hearing in a shirt-sleeve environment.

Soyuz

Just over two years later, in April 1967, the Russians began their long-running series of Soyuz (Russian for 'union') spacecraft missions. To date, over fifty Soyuz vehicles, of varying configurations, have been launched and the craft has achieved its original objective: to be a multi-purpose machine capable of supporting a space station as a personnel and cargo transporter, and of fulfilling independent operations. To this end, the early versions consisted of two compartments joined together by an airlock: one for launch and recovery (the command, or re-entry, module), the other for work in orbit (the orbital module). One, two, or three cosmonauts could be carried, and a docking unit could be fitted to allow two craft to join together. The craft was just over 8 m (26 ft) in length and 2.4 m (8 ft) in diameter at its widest point, giving a total volume of about $8.5\,m^3$ ($100\,ft^3$).

The flight of Soyuz 1 ended in tragedy when its parachute system failed to deploy correctly after re-entry, and Cosmonaut Vladimir Kopmarov was killed on impact. This event delayed the programme for eighteen months but then six Soyuz missions were flown within one year (Soyuz 3–8; Soyuz 2 was unmanned) and developed the ability to rendezvous and dock in space, and to carry out complicated manual tasks during EVA. Extensive medical monitoring had continued throughout these flights but apart from an altogether understandable increase in both heart rate and respiratory frequency during the launch and re-entry phases, and during EVA, no unexpected or unusual physiological responses were noted.

Soyuz 9, however, was an eighteen-day mission (the longest yet) specifically designed to study the effects of microgravity on many aspects of body function and to assess the value, if any, of various countermeasures against the expected cardiovascular and muscular deconditioning. Despite the use of techniques such as isometric exercises, tension straps, and chest expanders, the two cosmonauts were severely disabled on return to Earth by the phenomenon of orthostatic intolerance: i.e. a failure to react appropriately to the pooling of blood in the lower body on return to Earth which is manifest as a rise in heart rate, a fall in blood pressure, and a tendency to faint. These skills in space and experiences in physiology allowed the Russians to progress rapidly and well-prepared towards their orbital space station project: the Salyut (Russian for 'salute').

(It should be noted at this point that, although the successful manned exploration of the Moon is quite correctly attributed to the American Apollo missions, the Russians had progressed a long way on the road to a lunar landing. Indeed, Russian unmanned Luna spacecraft were the first to orbit the Moon (Luna 10 in 1966) and to soft land on the Moon (Luna 19, also in

1966). In addition, in what was a declared series of flights to precede a manned mission around the Moon, four Zond spacecraft (which were essentially Soyuz craft without the orbital module) were sent into lunar orbit, and then successfully recovered to Earth, in the period September 1968 (Zond 5) to October 1970 (Zond 8). Events, in the form of the spectacular American successes, obviously overtook the programme, which then went into abeyance.)

Salyut

Salyut space stations were designed to receive a series of crews sent up from Earth in Soyuz spacecraft, and were thus intended for long-term habitation. The early Salyuts were cylinders about 21 m (69 ft) in length, with a maximum diameter of nearly 4.2 m (14 ft), and each was equipped not only with a complete working environment but also with recreational and sleeping facilities for the crew. Power was generated on board by means of solar panels. Although the Soyuz 10 craft docked with the first space station (Salyut 1 was launched, unmanned, three days earlier) in April 1971 it was not until early in June that the station was first occupied by the three cosmonauts from Soyuz 11. Geophysical and biomedical research was carried out for the twenty-three days of the mission but, sadly, the crew of Cosmonauts Georgi Dobrovolsky, Viktor Patsayev, and Vladislav Volkov perished when their Soyuz command module failed to separate correctly from the orbital module, and the spacecraft underwent rapid decompression during the return to Earth. A further delay in the programme followed this setback and Soyuz 12 was not launched, on a proving flight, until September 1973: it had been modified to accept only two crew members, the third crew couch having been replaced by a life support unit, and there was a reversion to wearing full spacesuits instead of simple coveralls. In the meantime, Salyut 2 had also been launched but was apparently damaged at some stage and was never occupied.

Salyut 3, on the other hand, was launched in June 1974 and manned successfully for sixteen days that summer by the crew of Soyuz 14, who were able to enjoy its improved general facilities and more efficient life support systems, including enhanced solar panels. Salyut 3, which is believed to have been a military station, remained in orbit for a further six months but, apart from a failed docking manoeuvre by Soyuz 15, was not manned again. Salyut 4, with further improved solar panels capable of providing 4 kilowatts of power, was launched in December 1974 and was manned during 1975 by the crews of Soyuz 17 and 18B for thirty and sixty-three days respectively (Soyuz 16 had been a proving flight for the joint US/USSR Apollo-Soyuz Test Project (ASTP) and Soyuz 19 was the Russian craft used in this successful venture in July 1975). Salyut 5, another military station, was visited by three crews, those of Soyuz 21, 22, and 24, for a total of seventy-four days.

Throughout these missions of ever increasing duration, the Russians continued to investigate the physiological effects of microgravity and, where possible, to adopt active countermeasures to minimize the changes which occurred while weightless and to reduce the recovery period on return to Earth. To this end, extensive and regular exercise regimens were introduced, involving the use of sophisticated treadmills to allow walking, running, and jumping, and of weightlifting devices: all while wearing a special garment, the 'Pengvin (Penguin) Suit', which provided elastic resistance to certain movements so forcing the muscles involved to remain active. For several days before leaving the station, dietary measures such as increasing salt and fluid intake were added to the cosmonauts' daily round in order to reduce the period of orthostatic intolerance when back on Earth.

During the final stages of the Soyuz 21 mission, which eventually lasted forty-eight days, psychological problems were reported by the crew and described by *Isvestia* as 'sensory hunger which occurs when a human is cut off from the sights, sounds and smells to which he is accustomed'. This was one of the earliest descriptions of the ever present behavioural aspects of spaceflight; aspects which the Russians have always regarded as being of great importance. In the case of the Soyuz 21 crew, the solution to the problem of sensory isolation was to play music to the cosmonauts; and a similar philosophy persists in present Soyuz/Salyut/Mir missions in the form of family videos and tapes not only of music but also of rain falling, leaves rustling, bird song, and other earthly sounds. It is also of interest that the same mission, Soyuz 21, was terminated early when the crew was forced to abandon the Salyut as it filled with acrid fumes.

By 1977, a second generation of space stations had been developed to cope with the logistics of the prolonged periods of habitation now planned. This new improved Salyut had two docking points so that unmanned and dispensable cargo craft could re-supply the station at regular intervals. These so-called 'Progress' cargo vehicles were actually modified Soyuz craft and were capable of delivering 2,300 kg (5,000 lb) of stores on each journey: previously, the Salyuts and their visiting Soyuz transport craft could carry supplies for a maximum of only 120 days, each crew member requiring over 10 kg (22 lb) of consumables per day. The Progress craft were subsequently loaded with waste and redundant equipment from the space station before being disengaged and allowed to burn up on re-entry to the Earth's atmosphere.

Salyut 6 was the first of these larger stations and carried even more crew facilities, including full galley services, entertainment equipment, and a permanent shower. The craft interior was painted in pastel colours to enhance its homely appearance; and much effort was expended in reducing extraneous noise levels, a particularly bothersome aspect of earlier Salyut life. It was launched in 1977 and remained active until mid-1981. During this period fifteen Soyuz craft visited the station carrying a total of twenty-seven cosmonauts, including several from other Soviet bloc countries as part of the Interkosmos programme of peace and co-operation.

A policy of inhabiting the station with prime crews, who remained in orbit for long periods, while other spacecraft visited for shorter lengths of time, was adopted. Very long missions were completed by the men of the five prime crews: lasting 96, 140, 175, 185, and 75 days (one man, Cosmonaut Ryumin, was aboard for two of these flights). The practice of leaving the newest visiting Soyuz craft at the space station for the prime crew to use, and returning to Earth in the older vehicle, was also implemented by the short-term crews. In addition, there was a change in the Soyuz transport craft, Soyuz 40 being the last of the original vehicle type and launched in May 1981. During the year prior to this the new Soyuz T craft, basically a three-seater version of the old Soyuz but with on-board computers which allowed the crew more control over improved spacecraft systems, had been launched successfully on three occasions. The last of these (Soyuz T-4) ferried the final prime crew to Salyut 6.

Salyut 7 was launched in April 1982 and was essentially the same as its immediate predecessor. Salyut 7 is still in orbit and it, too, received prime crews at regular but increasingly extended intervals and, until the launch of the new Mir space station in 1986, was clearly being used by the Russians to establish year-round space station operations. Soyuz T-5 ferried the first prime crew to Salyut 7 in May 1982 and these two cosmonauts remained in orbit for 211 days. They were visited for short periods by a succession of other cosmonauts, including the first Frenchman in space (Jean-Loup Chrétien in Soyuz T-6, June 1982) and the second woman in space (Svetlana Savitskaya in Soyuz T-7, August 1982). In the latter's case, the interest in female physiology in space that started during Tereshkova's flight was continued. The next two-man prime crew arrived in June 1983 on board Soyuz T-9, the 'just visiting' Soyuz T-8 having failed to dock with the station two months earlier. In March of that year, however, a new unmanned Soviet space vehicle, the Cosmos 1443, had docked with the Salyut, so increasing the station's available work and living space, and providing other extra resources such as power. The space station capability was further enhanced when this module was successfully disengaged and returned to Earth, again unmanned but carrying material, in August 1983. The Soyuz T-9 crew remained on board the space station until November of that year, having received no visitors. A mission was attempted in the autumn but the spacecraft exploded during the launch sequence: the two cosmonauts on board survived because their escape rockets functioned correctly and carried the crew capsule away from the doomed Soyuz.

On 3 February 1984 the longest space mission thus far began with the launch of three cosmonauts in Soyuz T-10. This prime crew remained on board Salyut 7 for nearly 238 days, returning to Earth in October 1984. The crew included a doctor and, at that time, it became the declared policy of the

Russian programme that all long-duration missions would have a physician on board. Two short-stay crews, those of Soyuz T-11 and T-12, visited Salyut 7 during this record-breaking eight-month period. The former included an Indian cosmonaut and the latter Svetlana Savitskaya again. She was, therefore, the first woman to voyage into space for a second time, and on this occasion became the first to undertake EVA (Plate 3).

In early 1985 it was announced that Salyut 7 had completed its programme of work and the station was closed down, so prompting speculation that another Salyut was soon to be launched. In June, however, Soyuz T-13 was launched and its two-man crew set about rehabilitating Salyut 7 (which had suffered a solar panel malfunction and had frozen up). Following the successful re-activation, Soyuz T-14 visited in September 1985 and two of its three-man crew transferred to the station; the remaining crew member returned to Earth with one of the Soyuz T-13 cosmonauts, so effecting the first crew exchange since 1969.

A new space vehicle, the Cosmos 1686 Star Module, was also launched in September 1985 and docked with the Salyut at the position vacated by the returning Soyuz. This module was 13 m (43 ft) long and virtually doubled the station size. As did its Cosmos 1443 predecessor, the 1686 had a large returnable module at one end. The Soyuz T-14 mission came to an abrupt and premature end when, in November 1985, Cosmonaut Vasyutin became ill with what was originally thought to be an abdominal complaint which required surgery, and all three crew members returned to Earth. Subsequent information has suggested that a psychiatric illness was the real reason. This mission was the first to be aborted because of illness in space, and it is interesting to note that, despite the earlier declaration to carry a physician on long-duration missions, there was no doctor in the Soyuz T-14 crew. So, at the end of 1985, the Salyut 7 space station was once again operating in its unmanned mode. It was reactivated by the crew of Soyuz T-15/Mir in May 1986, who completed certain procedures necessarily omitted during the hurried evacuation of the Soyuz T-14 crew and carried out other tasks, including construction techniques during EVA, before returning to Earth in mid-July.

This extremely long Soyuz-Salyut programme has ensured that the Russians maintain an enormous advantage, in terms of time spent in space, over the Americans; but their objectives have been somewhat different. So far, just seventy-eight cosmonauts have achieved a total of nearly 6,000 man-days in space! Table 2.1 summarizes the Russian experience in space to the end of September 1988.

Mir and beyond

It has been quite clear for some time that the Russians intended to consolidate and expand their manned space station experience, and in February 1986 a

Table 2.1 Manned spaceflight: Russian experience

Programme	Period	Crew	Number of Flights	Number of Cosmonauts in Programme*	Total Flight Time (Days)	Total Man-days
Vostok	Apr 61 – Jun 63	1	6	6	15.9	15.9
Voskhod	Oct 64 – Mar 65	2 & 3	2	5	2.1	5.2
Soyuz/Salyut	Apr 67 – Nov 85	1, 2 or 3	50	66	1,842.6	4,074.1
Apollo–Soyuz Test Project	Jul 75	2	1	2	6.0	12.0
Soyuz/Mir	Mar 86 – Sep 88	2 or 3	6	16	758.0	1,823.0

Note: To September 88, 76 men (including 13 foreign nationals) and 2 women have spent a total of 5,927.2 man-days in space.
* Some have flown in more than one programme and on more than one mission in a programme.

17

large unmanned space station was launched. Christened Mir (the Russian for 'peace'), this third-generation station consists of an improved Salyut-like core, 15 m (49 ft) long and 4.15 m (13.6 ft) in diameter, with five docking points at one end, and a sixth at the other, each capable of receiving not only Soyuz-T, Progress, and Cosmos modules as before but also other Mir craft and even Salyut stations: the potential for a very large continuously manned orbiting laboratory has thus been taken a step closer to realization. Mir was activated in March 1986 by the two-man crew of Soyuz T-15: Cosmonauts Kizim and Solovyov, both of whom were veterans of the record 237-day mission of Soyuz T-10. Future inhabitants of Mir will travel in the new generation of upgraded Soyuz TM transport craft, the first of which was launched, unmanned, in May 1986; while the first manned mission (Soyuz TM-2) using this version conveyed a prime crew of two to Mir in February 1987. One of these cosmonauts, Alexander Laveikin, returned to Earth prematurely, with the visiting crew of Soyuz TM-3, five months later when he apparently developed an unspecified cardiac condition. The third member of the original TM-3 crew took Laveikin's place on Mir. The TM-2 mission finally ended in November 1987 when Cosmonaut Romanenko returned to Earth after a record 326 days in space. Ultimately, Mir has the capacity to accommodate twelve cosmonauts routinely and up to twenty occasionally. Later stations will probably consist of a spinning habitation module and a central, zero gravity, working environment. It is also likely that any deep space missions undertaken by the Russians or, as is conceivable, by international crews, will depart from such stations placed in Earth orbit.

Finally, it is known that the Russians have been developing a small, 'Shuttle-like' spacecraft of their own: the Kosmolyot. This rocket-powered, delta wing craft is launched from the back of a large booster after the combination has reached an altitude of about 18.6 miles (30 km). Indeed, scaled-down versions of these so-called Air-Launched Sortie Vehicles (ALSVs) are known to have flown already, albeit unmanned so far.

The American manned spaceflight programme

In the years following the end of the Second World War, the Americans were as actively involved in space research as the Russians and, although the latter were the first to launch a man into space, the former had sent five small rhesus monkeys aloft on separate V2 rocket flights by 1950. The animals did not survive the experience but the missions did serve to elucidate the requirements of life support systems and to point the way to the biomedical hazards ahead.

In 1958, largely as a response to the Russian demonstration of space technology (Sputnik), President Eisenhower established a new, civilian, organization – the National Aeronautics and Space Administration (NASA) – the purpose of which was to be 'the peaceful exploration of space'. Under the co-ordinating influence of NASA, the United States then continued its research into both unmanned and manned space exploration. There is no doubt that the launch of Gagarin into space by the Russians in the spring of 1961 once again severely shocked the Americans, who had until that time proceeded at an orderly pace in their own manned spaceflight programme. The supposed military threat from the East which manned spaceflight was assumed to pose prompted an enormous redoubling of American effort and, in May 1961, Project Mercury was coaxed into fruition. The project, first called MISS (Man In Space Soonest), had been inaugurated by Eisenhower in 1958 with the declared aim of launching a man into space, and recovering him safely, as soon as possible. The project was to be second only to national defence in importance, and the capability for prolonged manned missions in space was to be advanced at the same time. Thus, when the Russians launched Vostok 1, the Americans were not too far behind.

Throughout the mid to late 1950s, the Americans had undertaken a series of high-altitude balloon flights, including those of the Man High and Excelsior Projects. These programmes were used to assess the design of spacecraft life support systems (Man High) and of systems for escape at high altitude (Excelsior). The Man High series of flights culminated in a world record manned balloon ascent to an altitude of 101,516 feet (30,942 m) in 1957, and helped to prove the technology not only for life support but also for the means of monitoring both the systems and their users from the ground (i.e. radio-telemetry). Although the Man High balloon gondolas (cabins) were pressurized, their occupants were obliged to endure long periods of acute discomfort: David Simon's world record flight lasted over thirty-two hours (he was, incidentally, an air force doctor). Other equally discomforting balloon flights were carried out in unpressurized gondolas for Project Excelsior, during which the crew wore pressure suits and were asked to bale out at very high altitude to test the parachute systems. During the final high-altitude flight in 1961, to test spacesuits for Project Mercury, the extant altitude record for a balloon was established at 113,700 feet (34,665 m). The engineers and life scientists were then faced, as had been their Russian counterparts, with translating these and other, ground-based, achievements into a spacecraft system in which the needs of man in terms of an acceptable atmosphere (with adequate oxygen at adequate pressure) unpolluted by metabolic waste products, at an acceptable temperature and with provision of adequate nutrition, were constrained by the engineering needs of size, weight, energy supply, and reliability under extreme conditions of temperature and acceleration.

At the same time, the criteria for selection as an astronaut were being defined, although Eisenhower had decreed that, at least initially, all American candidates should be military test pilots. The immensely challenging physical and psychological procedures have only recently been relaxed with the advent of non-pilot, non-military crew members on board Shuttle flights.

Mercury

The Mercury spacecraft was essentially a conical tin can, but actually made of titanium, only just large enough to accommodate one man. Almost all of its control systems were fully automatic and the astronaut was, therefore, really an observer; he was, however, able to control spacecraft altitude and to fly the machine manually in the event of an emergency. If a pre-launch emergency arose, however, an emergency escape rocket mounted on top of a tower attached to the spacecraft would automatically carry the capsule away from the launch vehicle. After a successful launch, the escape rocket was jettisoned. The spacecraft also incorporated a de-orbiting retro-rocket, a retarding brake parachute for use after re-entry, and a recovery system involving water landing. The now familiar sight of frogmen surrounding an American spacecraft in the water after re-entry became standard procedure following the second Mercury mission. The flight of 'Liberty Bell 7' with the ill-fated Virgil (Gus) Grissom on board, nearly ended in tragedy in the sea when the capsule's escape hatch was blown prematurely and the spacecraft began to sink.

The first two Project Mercury missions, in May and July 1961, were sub-orbital flights, and lasted for just fifteen and sixteen minutes respectively. Although both Alan Shepard (the first American in space) and Grissom were fully instrumented and monitored, no significant physiological problems were noted. The third and fourth flights of Mercury, the first American orbital missions, both lasted almost five hours and again no physiological abnormalities were detected. The pilots of the Mercury 5 and 6 missions (nine and thirty-four hours respectively), however, were found to have orthostatic intolerance post-flight in that they were dizzy on standing up when back on Earth. These observations, along with those of weight loss and haemoconcentration (increased concentration of red blood cells as a consequence of reduced plasma volume), in part as a result of dehydration, formed the principal biomedical findings of Project Mercury. But the Project clearly achieved its principal aim: man could survive in space and return safely to Earth.

Gemini

Project Gemini was a natural progression of Project Mercury and its planning began just as the flights of Mercury were beginning in May 1961. Gemini was to be a series of two-man missions with the primary aim of establishing the capability for spaceflights of a duration at least as long as was required for a manned mission to the Moon. Secondary but no less important aims were to include perfection of techniques for rendezvous, docking, and extravehicular activity while in orbit, and for precise control of launch, re-entry, and recovery.

The Gemini spacecraft was a larger version of its predecessor and accommodated two crew members. Furthermore, the spacecraft systems were orientated towards primary control by the astronauts, rather than the fully automated systems of Mercury. The spacecraft comprised two modules. The first, the Adapter Module, was jettisoned just before re-entry and housed retro-rockets for use just prior to that event (or for use during launch emergencies) and thruster units for adjustment of orbital attitude. The second, the Re-entry Module, was itself made up of a parachute, radar, and docking section, an automatic re-entry attitude control section, and the crew cabin. The last was a truncated titanium cone, 1.9 m (6.2 ft) high and 2.29 m (7.5 ft) across at the base narrowing to 0.97 m (3.2 ft) at its top. Above each astronaut's couch was a windowed EVA hatch.

The first manned Gemini mission, Gemini 3, was launched in March 1965 with Astronauts Grissom and Young on board. The flight lasted for nearly five hours and involved extensive physiological monitoring. Gemini 3 was followed, during the next twenty months, by a further nine successful manned missions. Many notable firsts were achieved by the sixteen astronauts involved in the programme, including the first American EVA (Gemini 4) and the first American rendezvous and docking (Gemini 8 with an unmanned Agena spacecraft). The project achieved its operational objectives with outstanding success.

It was also vitally important from the medical standpoint. The four-, eight-, and fourteen-day missions of Gemini 4, 5, and 7 respectively demonstrated conclusively that there would be no physiological bar to the forthcoming lunar missions of Project Apollo, although the astronauts did become intensely bored and lethargic. Pre-flight, in-flight, and post-flight studies, particularly of cardiovascular function, had, however, confirmed the findings of Project Mercury. Thus, a loss of red cell mass was again seen, as was a universal susceptibility to orthostatic intolerance, to varying degrees, on return to Earth. A decline in exercise capacity was also demonstrated in flight, along with a loss of bone calcium and muscle nitrogen. It was clear that further elucidation of these changes, and especially of the time courses involved, would be an important aspect of any succeeding manned spaceflights. Additionally, the metabolic cost of EVA, in terms of oxygen need, was found to be higher than predicted. It is interesting to note, however, that no problems with motion sickness occurred during the Gemini missions, particularly since the Russian cosmonaut Titov had reported nausea during the second Soviet manned spaceflight four years earlier.

Apollo

Project Apollo also had its formal genesis in 1961 although it had first been mooted several years earlier as a programme for circumlunar flights. In 1961, however, President Kennedy altered the ultimate goal of Apollo to that of a

Moon landing and galvanized the project by calling for its accomplishment 'before this decade is out'. In accepting this challenge, it became clear that some new biomedical aspects would have to be addressed. These included the safety and medical well-being of the astronauts during such relatively prolonged missions: until Apollo, illness in flight had not been a major concern. The contamination of Earth by extraterrestrial organisms was also a novel potential hazard which prompted the development and implementation of quarantine and decontamination procedures until the risk had been defined. Finally, of course, the programme provided the opportunity for further study of the physiological effects of exposure to space.

The complete Apollo spacecraft consisted of a Command Module, a Service Module, and a Lunar Module. The Command Module (CM) was a conical pressurized cabin, about 3.3 m (11 ft) long and 3.9 m (12.8 ft) wide at its base, designed to accommodate three crew members in its central section. The couches on which the crew lay faced the apex of the cone where the display consoles were located, and five windows around the module provided an extensive view of the outside 'world'. The flight commander, who operated the flight controls, occupied the left-hand couch, with the CM pilot in the centre and the Lunar Module pilot in the right-hand couch. Finally, this section of the CM had two hatches: one was on the side and was the entry/exit hatch, and one was on the top and was the means by which transfer to and from the Lunar Module was accomplished. The forward section of the CM contained two reaction control engines and the recovery parachutes, while the rearward section housed fuel, gas, and water supplies, as well as a further ten reaction control engines.

The cylindrical Service Module (SM) was mounted below the CM and housed the main propulsion units together with most of the consumable supplies. It was about 7 m (23 ft) long and 3.9 m (12.8 ft) in diameter, and was jettisoned just before re-entry to the Earth's atmosphere.

The Lunar Module (LM) was designed to transport two astronauts to the Moon's surface and to return them to the orbiting CM. Thus it was a two-stage craft with an overall height of 7 m (23 ft) and a diameter of 3.3 m (11 ft). The whole unit was encased in an adapter module for aerodynamic reasons until the Apollo left Earth orbit on its way to the Moon. The lower, descent stage was the unmanned portion and contained the descent engine and its propellants, which provided the deceleration required for a lunar landing. It also housed various items of support material such as sample containers, television equipment, and, in later missions, the Apollo Lunar Surface Experiment Package (ALSEP) and the Lunar Roving Vehicle (LRV). The ascent stage, which used the descent stage as its launch pad when leaving the Moon, itself comprised three sections: the pressurized crew compartment, a pressurized mid-section, and an unpressurized aft equipment bay which housed the ascent engine. The ascent stage provided the crew with living quarters while on the Moon and acted as their operational base. Once it had

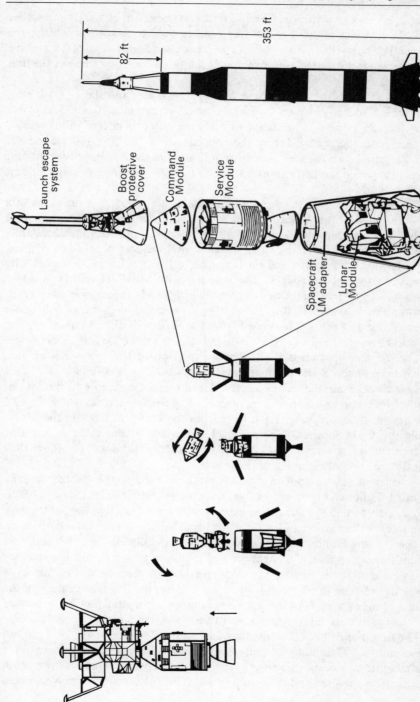

Launch escape system

Boost protective cover

Command Module

Service Module

Spacecraft LM adapter

Lunar Module

Figure 2.1 Configuration of the Apollo spacecraft

successfully docked with the SM and returned the astronauts to their mother ship, the LM was allowed to fall free back to the Moon's surface. Since the LM could only fly in the vacuum of space, and was unable to survive unprotected in the Earth's atmosphere, it has often been described as the first true spacecraft.

Finally, in addition to the three principal modules and the LM adapter, there was a rocket-powered Launch Escape System, in a 10 m (33 ft) tower mounted above the CM, designed to carry the latter to safety in the event of an emergency occurring during the countdown or the first 100 seconds of flight. Figure 2.1 shows the launch configuration of the Apollo spacecraft and the way in which the LM was extracted by the CM from its housing while en route to the Moon.

The configuration of the Apollo modules, as described, was the result of a decision to use the lunar-orbit rendezvous technique as the means of achieving a manned Moon landing within the constraints of funds available, safety, technological ability, and, in view of President Kennedy's dictum, time. This technique involved the launch of a spacecraft into Earth orbit by means of a single large rocket – the Saturn V with its I-C, II, and IV-B stages – and then into lunar trajectory by means of the third-stage rocket – the IV-B. During the journey to the Moon, the Command and Service Module combination (CSM) was separated from the Saturn IV-B and turned around 180° to allow the top of the CM to dock with the top of the LM ascent stage. The CSM then extracted the LM from the Saturn IV-B, which was then discarded in such a way that, in later missions, it would impact the lunar surface to calibrate the seismometers placed there. Once close to the Moon, the Apollo complex's own rockets placed it in lunar orbit, from which the LM component departed, after separating from the CSM, to land on the Moon while the CSM mother ship remained aloft. The ascent stage of the LM carried the lunar visitors back to the CM (the rendezvous), to which they transferred for the return leg of the journey.

A series of six unmanned Apollo missions during 1966–8 preceded the manned flights and were used to validate the launch vehicles (Saturn I-B, I, and V rockets) and the many new aspects of spacecraft design needed to carry men safely to the Moon and back. Thus, these missions included qualification of the Saturn launch and stage-separation capabilities, the SM and LM propulsion systems, the navigation and guidance systems, and the re-entry heat shields. These flights, in turn, had been preceded by the Surveyor missions designed to establish advanced soft lunar landing techniques and the nature of many aspects of the lunar environment. Potential landing sites were also surveyed during three unmanned Lunar Orbiter flights.

The first manned Apollo mission was scheduled for February 1967 but, on the evening of 27 January of that year, while the first full dress rehearsal for the mission was being carried out, the oxygen-rich atmosphere of the cabin was somehow ignited and the resulting flash fire killed all three crew members within seconds. This test, designated Test Apollo 204, was re-named Apollo 1 in honour of the crew: Virgil Grissom (one of the 'Original Seven', Commander of Mercury 2 and Gemini 3), Edward White (the first American to 'walk' in space in Gemini 4), and Roger Chaffee, whose first flight it was to be. The cause of the accident has never been fully elucidated but it resulted in a delay to the programme while alterations were made to the composition of the spacecraft atmosphere and to the materials used in certain components. Thus, atmospheric composition was changed, for all subsequent launches, from one of 100 per cent oxygen to one which contained approximately 40 per cent nitrogen. The nitrogen was gradually replaced by oxygen once the high-risk launch period was over. In addition, inflammable rubber and plastic components, including boot soles, belts, and hoses, were replaced by, or coated with, newly developed non-flammable fluorinated hydrocarbon compounds. The tragedy also resulted in a major curtailment of the planned medical investigation programme as all effort was re-directed solely towards the primary task. It was decided, however, that henceforth all three crew members would be physiologically monitored during launch, instead of just one member as originally intended.

Apollo 7, with Astronauts Schirra, Eisele, and Cunningham on board, was launched on 11 October 1968. This first manned Apollo mission lasted nearly eleven days and was an Earth orbital test flight. It succeeded in its primary aim of proving both crew and spacecraft performance over the period it would take for a lunar landing mission. There was very little biomedical investigation involved in the Apollo 7 flight but some spacecraft operations were hindered by the development of upper respiratory tract infections (head colds) in the crew. Apollo 8 was launched two months later and was the first manned lunar orbital flight. Lunar orbit was achieved on Christmas Eve 1968 and the crew of Borman, Lovell, and Anders flew around the Moon ten times before the SM engines were fired to start the return to Earth. Sixty-three hours later, Apollo 8 splashed down in the Pacific Ocean after six spectacularly successful days in space.

In March 1969, the crew of Apollo 9 (McDivott, Scott, and Schweickart) completed the penultimate proving flight of the Apollo/Saturn complex. Once again, the crew suffered from head colds during the mission; and Schweickart was forced to delay a planned space walk because of nausea and vomiting during the early days of the flight. By the fourth day, however, he was well enough to leave the LM, to which he and McDivott had transferred, and to stand on its porch: so qualifying the Apollo Portable Life Support System (PLSS) backpack. The next day, McDivott and Schweickart separated their Lunar Module from the CM and for several hours flew at distances of up to 190 nautical miles (352 km) from the mother ship. They subsequently separated the ascent from the descent stage and successfully re-docked with the CM.

Apollo 10 was the last planned lunar orbital mission and was a full dress rehearsal for a lunar landing. The crew, Astronauts Stafford, Young, and Cernan, was launched on 18 May 1969 and the flight lasted eight days. On the fourth day, Stafford and Cernan undocked the LM and flew it to within 50,000 feet (15,480 m) of the lunar surface before re-establishing lunar orbit and returning to the CM for the journey back to Earth.

Apollo 11 was launched from the Kennedy Space Centre in Florida on 16 July 1969. Its flight was to last just eight days but, in that time, two of its crew had set foot upon another world and manned spaceflight had come of age. Furthermore, by the wonder of radio-telecommunication, millions of people left back on Earth were able to witness and enjoy this epoch-making achievement. Lunar orbit was attained by 19 July and the next day, Sunday, Astronauts Neil Armstrong and Edwin (Buzz) Aldrin entered their Lunar Module, named by them 'Eagle'. Shortly afterwards, the LM was separated from the Command Module and descended towards the lunar surface. Armstrong flew the final stages of the descent manually to avoid the large rocks unexpectedly strewn about the Sea of Tranquillity, and 500 million people on Earth heard him say 'the Eagle has landed'. Six hours later Armstrong became the first human being to stand on another planet, and Aldrin followed after twenty minutes. The two men remained on the Moon's surface for two and a half hours during which time they set up several scientific experiments (including a passive seismometer, a solar wind analysis device, optical reflectors for laser experiments from Earth, and an Earth–Moon communications link) and collected about 21 kg (46 lb) of lunar rock and soil samples. The LM ascent stage was fired faultlessly several hours later and it docked with the CM three and a half hours after leaving the Moon. The return to Earth was likewise uneventful, although the crew was pre-occupied during this time with preparations for quarantine. The three men were obliged to spend twenty-one days in isolation undergoing medical and biological tests (see Chapter 9) before the programme was ended and they could enjoy unhindered their fully deserved acclaim.

So, with six months to spare, NASA had successfully met the challenge laid down by President Kennedy. But a period of retrenchment was to follow, during which severe budgetary cutbacks were imposed by a now disinterested United States administration. A modified programme of lunar landing missions was able to proceed, however, and a further six Apollo missions were undertaken in the ensuing three and a half years.

Apollo 12, launched in November 1969, succeeded in its primary aim of achieving a pre-determined point landing on the Moon, despite being struck twice by lightning shortly after take-off. The crew of this second LM, Conrad and Bean, spent over thirty-one hours on the Moon, much of it working on the lunar surface. During their two EVAs, they placed a second ALSEP and collected 35 kg (75 lb) of lunar material. The safe return of the entire crew to Earth was again followed by a twenty-one-day incarceration in quarantine;

but again no biological threat was isolated.

In the spring of 1970, Apollo 13 was launched on what was supposed to be a purely geological research mission. It had already been the subject of intense medical interest when the nominated CM pilot, Thomas Mattingly, was found to have been exposed to rubella (German measles) contracted by a back-up crew member with whom he had been training. Mattingly did not have protective antibodies against rubella in his blood and, although he ultimately failed to develop the illness, no risk could be accepted; his place on the ill-fated flight was taken by John Swigert. The spacecraft was about four days into its flight, and well on the way to the Moon, when an electrical short circuit in one of the SM oxygen tanks caused its rupture when temperature and pressure rose within it. The explosion disrupted the entire bay and the loss of oxygen, used both for the production of power and for breathing, presented an acute emergency. Fortunately, the LM and its support systems were undamaged and the crew rapidly transferred to this vehicle. Even then, however, the supplies available were sufficient for only about thirty-eight hours: the obligatory loop around the Moon and the return to Earth would take at least ninety hours. Emergency procedures and equipment, including, most importantly, adaptation of the CM lithium hydroxide cells to eliminate expired carbon dioxide from the LM, were devised by workers on the ground and improvised by the astronauts. By powering down the craft, vital supplies were conserved for the necessary time. The LM became extremely cold and uncomfortable as a result but, although one crew member developed a urinary tract infection, the three men were in reasonably good health on their return to Earth. They had abandoned their life-saving LM just one hour before entering the Earth's atmosphere.

No further manned missions were undertaken that year and Apollo 14 was finally launched on 31 January 1971, after weather had caused a short delay. The mission was commanded by America's first man in space, Alan Shepard, and, together with LM pilot Edgar Mitchell, he spent over thirty-three hours on the lunar surface. Another ALSEP was deployed and 43 kg (95 lb) of lunar material collected. One foray to collect rocks was abandoned, however, when both astronauts began to tire seriously and their heart rates increased to over 125 beats per minute. Apart from this, the mission was unremarkable from the medical point of view, perhaps at least in part because very rigorous pre-flight procedures were adopted to prevent a repeat of the Apollo 13 rubella problem. Crew contacts were restricted to wives and 150 essential personnel, and the buildings used were fitted with air filtration plants. The three-week post-flight quarantine period was again imposed but thereafter the programme was terminated, the Apollo 14 crew being the last to use the quarantine facilities.

The Apollo 15 mission, in July 1971, advanced the scientific exploration of the Moon still further. The LM was modified to double the potential duration of its sojourn on the lunar surface and the PLSS was also modified to prolong its usage without re-charging from four to five hours to seven to eight hours. The mission was the first to use the Lunar Roving Vehicle (LRV) and Astronauts Scott and Irwin spent more than nineteen hours exploring the Apennine mountain base, and a total of sixty-seven hours on the Moon.

If the Apollo 14 mission was without major medical incident, the crew of Apollo 15 redressed the balance. Potentially serious cardiovascular responses, in the form of cardiac dysrhythmias (disturbances of cardiac conduction resulting in premature heart beats and/or an irregular rhythm), were seen during lunar surface activity and during the return to Earth. The precise cause of these disturbances has never been identified although they were believed related to dietary potassium deficiency before take-off and to the very tiring workloads imposed during EVA. One affected crew member subsequently developed previously undetected coronary artery disease, and his dys-rhythmia may have been an indication of this at a sub-clinical stage. The crew of Apollo 15 also took three to four weeks, longer than any other American crew, to recover pre-flight levels of cardiovascular and exercise tolerance on return. A similar anomalous response had been noted by the Russians in the crew of Soyuz 9.

The crews of the final two Apollo lunar missions, Apollo 16 in April 1972 and Apollo 17 in December 1972, were subjected to strict pre-flight health programmes which, among other things, ensured that potassium balance was normal. They also took sedatives during the missions to ensure that adequate rest was taken to complete their scheduled lunar activities. The lunar explorations (Plate 2), which were again largely geological (204 kg (450 lb) of lunar material were collected) and geophysical (two more ALSEPs were deployed), were operationally and physiologically successful with no manifest medical problems.

With the successful return of Apollo 17, the Lunar Landing Program came to an end. Twenty-nine astronauts had flown on eleven manned missions, and a total of just over 600 man-hours on the surface of the Moon had been achieved by the twelve men privileged to land there. Medically, the vital importance of detailed pre-flight preparation was indisputably proven, while the necessary, but fortunately negative, investigations while quarantined post-flight had revealed no threat of contamination to the Earth by its Moon. The physiological findings of the Mercury, Gemini, and indeed Russian programmes had been confirmed: once more, when compared with pre-flight levels, there was post-flight dehydration, decreased orthostatic and exercise tolerance, decreased red cell mass, and less than desirable food consumption. To this list was now added disturbance of vestibular function: the nausea and vomiting experienced by many of the Apollo astronauts and which, in one instance, was so severe as to postpone part of the flight plan. Space Motion Sickness was now as familiar to the Americans as it had been to the Russians since 1961.

Skylab

Project Skylab, originally called the Apollo Applications Program, was a logical and important progression from the Mercury, Gemini, and Apollo adventures. The three principal objectives of the project were the study of man, the Earth, and the Sun. Skylab itself was constructed within the shell of a Saturn IV-B rocket and as such was relatively enormous. This Orbital Workshop (OWS) was cylindrical in shape, being nearly 14.6 m (48 ft) long and 6.7 m (22 ft) in diameter, and offered over 294 m^3 (10,383 ft^3) of living and working space: by contrast, the Apollo CM provided 5.95 m^3 (210 ft^3) and the Mercury cabin a mere 1 m^3 (35.3 ft^3) of habitable space.

Internally, Skylab was divided into two main sections: an upper working area where large-scale experiments could be undertaken and a lower living compartment with facilities for minor experiments and for eating, sleeping, relaxation, and personal hygiene. The Skylab laboratories were well-equipped for the study of astronomy, bacteriology, biology, botany, geology, physics, and zoology, as well as physiology: the fully-laden craft weighed over 34,500 kg (76,000 lb) on Earth. Power was provided by deployment of solar panels once in orbit, and the full configuration was completed by the addition of an Airlock Module and Multiple Docking Adapter mounted at one end.

This complex – designated Skylab 1 – was placed unmanned into Earth orbit by a Saturn V rocket on 14 May 1973. Eleven days later, Astronauts Charles Conrad, Paul Weitz, and Dr Joseph Kerwin (who was the first American physician to fly in space) were launched in a conventional Apollo CSM to man the space station as the complement of the Skylab 2 mission. Before docking, Conrad and Kerwin were obliged to spend several hours outside the Skylab effecting repairs to a damaged micrometeoroid shield (which was causing excessive thermal strain on the station) and releasing a solar panel which had failed to deploy. The entire programme would almost certainly have been lost had these efforts failed. The Skylab 2 mission lasted for twenty-eight days and was followed by the fifty-nine day mission of Skylab 3 (launched on 28 July 1973) and the eighty-four day mission of Skylab 4 (launched on 16 November 1973). The space station remained unoccupied from the end of the Skylab 4 mission in the spring of 1974 and, despite efforts to prevent the decay of its orbit, it eventually burnt up in spectacular fashion as it fell to Earth in July 1979.

The 171 days of Skylab manned operations provided a vast amount of scientific data covering many disciplines and that concerned with physiological adaptation to the environment of microgravity forms a large contribution to Chapter 11 of this book. Briefly, however, the Life Sciences Program devised for Project Skylab concentrated on six main areas of concern, as identified during previous projects:

- changes affecting the neuro-vestibular system,
- changes affecting the cardiovascular system,

29

- changes affecting fluid and electrolyte balance (salts, acids, and bases),
- changes affecting haematology (the study of blood),
- changes affecting the musculo-skeletal system,
- changes affecting the central nervous system.

Extensive experimentation into all these areas was undertaken in an attempt to clarify the changes seen, monitor their time scales, and to provide recommendations for the future support and study of long-term space missions.

In addition to the scientific research aspects of Skylab, the support of men in space for relatively long periods of time required other medical inputs of an operational/technical nature. These inputs, which are discussed in detail in Chapters 3, 6, 7, and 9, were also based on experience gained during earlier programmes and included design, function and monitoring of:

- the environmental control system,
- the food supply system,
- the waste management system,
- the in-flight medical support system.

Finally, to these long lists of biomedical relevance must be added the considerable ground-based effort required to select, train, and assess the chosen astronauts before, during, and after the flights themselves. These aspects are discussed in Chapter 10.

Apollo-Soyuz Test Project

The Apollo spacecraft was to make its final appearance as one half of the joint American-Russian Apollo-Soyuz Test Project (ASTP) in the summer of 1975. The ASTP was conceived as an exercise in international co-operation with the principal aim of demonstrating a mutual rescue capability for in-flight emergencies. In this it succeeded admirably when, at noon on 17 July, a conventional Apollo craft (Apollo 18) docked with a conventional Soyuz craft (Soyuz 19) and the two crews were subsequently able to visit each other's vehicles. A special in-flight docking module was constructed to enable the crews to transfer from one craft to the other, and to allow appropriate adjustment of cabin pressure to be made from the markedly different levels usually adopted by each nation.

The spacecraft remained together for two days, during which five joint experiments were conducted, before separating and returning to Earth. The return of the Apollo crew was somewhat marred by the inadvertent release of toxic gases, principally nitrogen tetroxide, into the cabin at an altitude of 24,000 feet (7,317 m). All three crew members developed a chemical pneumonitis (inflammation of the lungs) despite donning their oxygen masks rapidly; indeed, one crew member was rendered unconscious for a short time.

No permanent damage resulted but the requirement for intensive hospital treatment curtailed the routine post-flight medical investigations. Some valuable data were gathered, however, concerning skeletal muscle function and the speed of tendon reflexes.

A final point of medical interest in the ASTP was the inclusion of Astronaut Donald (Deke) Slayton as commander of the Apollo. Slayton was one of America's 'Original Seven' astronauts chosen for the Mercury programme but he was declared unfit to fly in 1961 following the reassessment of a cardiac dysrhythmia: it had taken him those fourteen years to re-establish his medical fitness to the satisfaction of the NASA project managers and doctors!

Space Transportation System

The American manned space programme was quiescent for nearly six years following the ASTP and it was not until 12 April 1981 that the re-usable Space Shuttle, the principal component of the Space Transportation System (STS), undertook its maiden flight. The STS was conceived as a means of providing a working space vehicle, capable of recovery intact after its flight and of being re-launched after a short turn around servicing period, so saving capital costs. It consists of three main components: a large External Fuel Tank, two Solid Rocket Boosters (SRBs), one on each side, and the Shuttle or Orbiter itself. The Orbiter is launched vertically, with its crew horizontally orientated as in all previous programmes. Over seven million pounds of thrust are imparted by the main Orbiter engine, fuelled from the External Tank, in combination with the SRBs for additional initial thrust. Once the SRBs have exhausted their fuel they are detached from the External Tank and are recovered to Earth by parachute for further use. The empty External Tank is jettisoned by the Orbiter shortly after the SRBs detach and is allowed to burn up on re-entry to the Earth's atmosphere: it is therefore the only expendable component of the STS. The Orbiter is recovered to Earth by adopting a conventional aircraft attitude, with wings horizontal, after re-entry; the pilot executing an unpowered approach and landing in the manner of a glider.

The Orbiter is about 37 m (122 ft) long with a fuselage width of about 7 m (22 ft): approximately the size of a DC9 aircraft. Its wing-span is 24 m (78 ft) and it weighs about 68,000 kg (150,000 lb) on Earth. Internally, the pressurized cabin has a total habitable volume of 71 m^3 (2,507 ft^3) with accommodation for a crew of from two to eight people in normal use, and for a total of ten in an emergency.

The cabin is divided into three levels. The upper level is the flight deck with seating for four, and it contains the controls and displays for the Orbiter itself and for its payload. The commander and pilot sit in front with a mission specialist, responsible for the co-ordination of scientific objectives, and a payload specialist just behind. The mid-deck provides the seating area for the other mission and payload specialists together with eating, sleeping, and toilet facilities; while the lower deck, reached by removing the mid-deck floor panels, houses the environmental control system and other equipment.

Once in orbit, all but the commander's and pilot's seats are dismantled and stowed until required for re-entry. The large (4.5×18 m (14.8×59 ft)) payload or cargo bay can be entered via an airlock in the rear of the pressure cabin and it is this area which gives the Orbiter its versatility, since it provides the location for the various payload configurations, including the principal experimental facility for the STS: the European Space Agency's (ESA) Spacelab.

Spacelab can vary in its specification according to the needs of each mission. Thus, it may consist of one or more pressurized modules, to which the crew can transfer via a tunnel from the mid-deck and in which various experimental racks can be mounted. The rack concept allows the construction of mission-specific experiments so that, for example, an Anthrorack exists for the study of human physiology and a Biorack for general biological studies. Alternatively, Spacelab can be composed entirely of unpressurized pallets, which again mount experiments in racks but this time for direct exposure to, and study of, the environment of space. Such experiments would be controlled from within the Orbiter's cabin by the payload specialists. A combination of modules and pallets is also possible. Other uses for the payload bay have included the transport and release of satellites, the rescue and repair of satellites, and of course there have been classified military payloads.

The first Orbiter was named Enterprise and made its debut in 1976. It was a test vehicle for launch pad and flight assessments, and the following year saw a number of test flights of Enterprise mounted on top of a modified Boeing 747 aircraft. This phase ended with a series of five free flights after release from the 747 at 24,000 feet (7,317 m) to prove the approach and landing capability. The first manned mission into space for the Shuttle was the Orbital Test Flight (OTF) of STS 1 and took place just over three years later in April 1981, when Astronauts Crippen and Young, the latter undertaking his fifth spaceflight, successfully returned their craft Columbia to Earth in a re-usable condition following two days in orbit. Columbia was used for the next four flights; STS 2, 3, and 4 were further OTFs designed to qualify the system, while STS 5 (launched in November 1982) was the first operational Orbiter mission and had a crew of four. Since then, despite a considerable number of technical difficulties, Orbiters were sent into space at increasingly frequent intervals and 1985 saw nine launches.

In January 1986, however, the Shuttle programme suffered a major and tragic setback when the Orbiter Challenger, on its tenth flight, suffered a failure of a seal in one of the SRBs and exploded just seventy-three seconds after take-off, killing all seven on board. The crew consisted of five men and

Table 2.2 Manned spaceflight: American experience

Programme	Period	Crew	Number of Flights	Number of Astronauts in Programme*	Total Flight Time (Days)	Total Man-days
Mercury	May 61 – May 63	1	2 sub-orbital 4 orbital	2 4	2.2	2.2
Gemini	Mar 65 – Nov 66	2	10	16	40.4	80.8
Apollo	Oct 68 – Dec 72	3	11 (6 lunar landings)	29	103.8	311.4
Skylab	May 73 – Feb 74	3	3	9	171.0	513.0
Apollo-Soyuz Test Project	Jul 75	3	1	3	9.0	27.0
Space Transportation System (STS)	Apr 81 – Jan 86	2–8	25	94	152.5	847.0

Note: To August 88, 123 men (including 8 foreign nationals) and 8 women have spent a total of 1781.4 man-days in space.
* Some have flown in more than one programme and on more than one mission in a programme.

33

two women: Francis Scobee (Commander), Michael Smith (Pilot), Ellison Ozinuka, Ronald McNair, and Judith Resnik (Mission Specialists), and Gregory Jarvis and Christa McAuliffe (Payload Specialists). Mrs McAuliffe was a social sciences teacher and was the first private American citizen to travel into space.

Despite this terribly sad event, the twenty-four successful missions to date have accomplished a number of very impressive 'firsts' for the American programme. STS 7 (June 1983) carried the first American woman, Sally Ride, into space; and several more women, including doctors, followed her, although all were mission or payload specialists rather than pilots (the 'true' astronauts). Kathryn Sullivan, on board STS 13 (October 1984), was the first American woman to undertake EVA. Numerous satellites, sponsored by many different countries, have been released from the Orbiter's payload bay and despatched into orbit, while others have been repaired in space and yet more have been retrieved for repair back on Earth. Spacelab has been flown four times (on STS 9, 17, 19, and 22) and the results of its many experiments in all fields of science have been immensely fruitful. Seven men and one woman made up the crew of the most recent Spacelab mission (STS 22 sponsored by West Germany and launched at the end of October 1985), so becoming the largest single crew in space so far. And two partially classified military missions have been accomplished (STS 15 and 21). The loss of Challenger (maiden flight: STS 6 in April 1983), however, has reduced the Orbiter fleet to three – Columbia (maiden flight: STS 1 in April 1981), Discovery (maiden flight: STS 12 in August 1984), and Atlantis (maiden flight: STS 21 in October 1985) – and this will inevitably disrupt the planning of future missions.

With the increasing pace of Shuttle launches, and the large numbers of astronauts carried on each mission, the American experience, in terms of time spent in space, has been growing relentlessly: over 120 astronauts have now achieved a total of nearly 1,800 man-days in space, and Table 2.2 summarizes the American experience to summer 1988. Now that the cause of the explosion which destroyed the twenty-fifth Shuttle mission in the twenty-fifth year of manned spaceflight has been fully investigated, and it has been determined that, with modifications to the fateful seals, such an event is unlikely to happen again, missions have been scheduled to begin once more late in 1988.

It is to be hoped that the STS programme will then blossom as the venture becomes commercially viable, for it was always intended that a Shuttle would be launched once every two weeks. Furthermore, with the announcement by President Reagan in early 1985 of a formal commitment to a Space Station by the 1990s, the STS should clearly have a vital role to play in the construction, manning, and replenishment of this new project. Other space vehicles are also on the drawing boards of America and even some of her allies. Many will have military applications, such as Air-Launched Sortie

Vehicles (ALSVs) and Spaceplane, but others, including perhaps the British HOTOL craft, the French-inspired Hermes spaceplane, and the European Space Agency's Columbus project, will advance the 'peaceful exploration of space' 'for free men [to] fully share'.

Part two

Putting Man into Space

When Werner von Braun perfected the use of rockets during the later stages of the Second World War, the way was open to realize the long-dreamed of ambition to travel into space. As we have seen, in the years following the end of the Second World War, the technology needed for space travel by man was developed both in the United States and in the Soviet Union. By 1950, primates had been launched into space on board V2 rockets and important lessons were being learnt: particularly the need for reliable and effective life support systems. There were scientists, however, who felt that space travel by man was too hazardous and would never be accomplished; and, even if it was, that the hostility of the space environment would preclude the safe return of the voyagers.

The chapters in this part describe the elements of this hostile environment and the measures needed to combat them in order that manned space travel, of any duration, can take place. The elements include the threat posed by the lack of atmosphere beyond the very thin shell around our Earth, by the sheer power of the rockets needed to carry men and equipment beyond the pull of Earth's gravity, and by the exposure to risks such as radiation and micrometeoroids, against which our atmosphere normally guards us. In addition, there are certain other basic needs to which attention must be paid if man is to stay in space for even a relatively short time. Thus, Chapter 6 deals with the problems of temperature and humidity control, and Chapter 7 with the supply of food and water, the management of human waste products, and aspects of personal hygiene. The provision of adequate clothing, including spacesuits and the problems of mobility, is discussed next, while the penultimate chapter addresses the health care required before (including medical selection), during, and after space missions. The final chapter in this part discusses some other aspects of selection and the biomedical training procedures which have produced the men and women who are not only able but also willing to withstand the undoubted rigours of space travel.

Chapter three

Pressure and density

The problems

Hypoxia

Man is adapted to life on the surface of the Earth where the total atmospheric pressure at sea level is normally about 760 mmHg (101.3 kPa) and the composition of the air he breathes is the familiar 21 per cent oxygen and 79 per cent nitrogen. The use of percentages, although common, is not helpful when considering human physiology, and this is especially so when at altitude. The total pressure exerted by a mixture of gases is the sum of the individual pressures exerted by each component of that mixture. Thus, at sea level, in the simple case described, oxygen is exerting a pressure of 21 per cent of 760 mmHg, i.e. 159.6 mmHg (21.3 kPa), while nitrogen is exerting a pressure of 79 per cent of 760 mmHg, i.e. 600.4 mmHg (79.7 kPa). For various reasons, including the addition of water vapour and carbon dioxide, the partial pressure of oxygen has fallen to about 103 mmHg (13.7 kPa) by the time air has reached the terminal air-sacs (alveoli) in the lungs of healthy individuals at sea level. It is this level of partial pressure, or driving force, of oxygen which is necessary to keep the most distant body cells alive. If, for any reason, the partial pressure of oxygen in the lungs falls, then the body is at risk from lack of oxygen; that is, from hypoxia. At sea level, such falls may be seen in many disease states, particularly those affecting the respiratory system (e.g. emphysema and chronic bronchitis), the cardiovascular system (heart and blood vessels) (e.g. ischaemic heart disease), or the blood itself (e.g. anaemias). Falls are also seen, even in healthy people, whenever they are exposed to high altitude.

Hypoxia is one of the most serious hazards facing anybody who flies, including astronauts. Its effects are seen with increasing rapidity and severity the greater the altitude to which the victim is exposed. Thus, for example, acute exposure to an altitude of 25,000 feet (7,620 m) will produce severe symptoms and signs within three to five minutes, with unconsciousness and death supervening shortly thereafter. The early symptoms and signs resemble those of alcoholic intoxication with euphoria, personality change, loss of self-criticism, lack of insight, and loss of judgement. It is these features which account for the great danger of hypoxia, since they are seldom recognized for what they are by the victim. Later, there is progressive loss of mental and muscular co-ordination, with lightheadedness, feelings of dissociation, jerky limb movements, and slurred speech. These muscular (motor) signs are accompanied by sensory loss, with numbness and paraesthesiae (pins and needles) of the hands, feet, and face. Vision is also degraded, with a decline in visual acuity and loss of peripheral visual fields (so-called 'tunnelling' of vision). Cyanosis (blueness) of the nail beds, ear lobes, and lips can be seen as the condition develops. At altitudes greater than 45,000 feet (13,716 m), unconsciousness develops in fifteen to twenty seconds with death following four minutes or so later.

Many of the overt symptoms and signs of hypoxia are attributable to just one aspect of the body's reaction to lack of oxygen: that is, to the respiratory response. To protect itself against a lack of oxygen, it is entirely predictable that the brain would attempt to increase the available air supply. Just as there is a reflex increase in the rate and depth of respiration during exercise to increase oxygen delivery, so too is there a reflex increase during times of reduced oxygen partial pressure. In the latter case, however, such increased respiratory effort can do very little to counteract the situation since oxygen is simply not present at sufficient pressure. Increasing the respiratory frequency and depth serves only to affect the other side of the metabolic equation: the elimination of the end-product of metabolism – carbon dioxide.

Although carbon dioxide is a waste product, and as such is expired with each breath, it also plays an extremely important role in the maintenance of the body's milieu (internal environment). Specifically, carbon dioxide in combination with water (that is, as carbonic acid) is responsible for the control of body acidity. Even slight alterations in the concentration of acid in the body will cause widespread effects. Thus, when the body responds to hypoxia by breathing more deeply and more quickly too much carbon dioxide is expired, the level of carbonic acid falls, and the body becomes alkaline. This alkalinity, throughout the body but especially in the brain, produces dizziness, lightheadedness, feelings of dissociation and anxiety, parasthesiae, and, if the condition progresses, muscle spasms (tetany). These are the symptoms and signs of hyperventilation (the definition of which is, therefore, breathing in excess of the body's needs to eliminate carbon dioxide) and they closely resemble the features of hypoxia already described. There are, however, some other causes of hyperventilation in flight besides hypoxia. The most significant of these is emotional stress or anxiety, and indeed this cause can produce the syndrome of hyperventilation even on the ground. So hyperventilation, while not strictly a problem of ascent to altitude, will certainly be evoked should hypoxia develop and may occur in its own right at times of stress. As spaceflight is clearly a stressful occupation, selection of

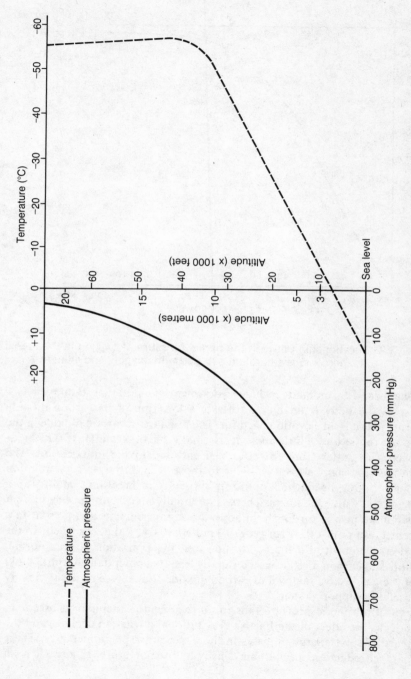

Figure 3.1 Relationship between total atmospheric pressure and altitude, and between total atmospheric pressure and temperature

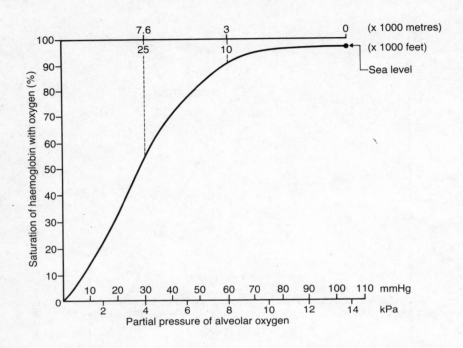

Figure 3.2 Relationship between the partial pressure of oxygen in the lungs, the percentage saturation of blood with oxygen, and altitude

astronauts and cosmonauts includes assessment of emotional stability.

Just as the body is unable to tolerate a lowering of its carbon dioxide content, so too is an elevation harmful. The pressure of carbon dioxide in the blood has to rise only a little above its normal value of 40 mmHg (5.3 kPa) for the victim to suffer from the effects of carbon dioxide intoxication. The symptoms and signs are very similar to those seen during hyperventilation but, in addition, headache, confusion, difficulty in breathing, and a rapid heart rate may all be features. There is essentially no carbon dioxide present in the air we inspire on Earth: in an outdoor environment it is present in a concentration of just 0.04 per cent (0.3 mmHg (0.04 kPa) at sea level). In the closed environment of a spacecraft, however, the expiration of even normal quantities of carbon dioxide would rapidly lead to a potentially harmful build up of the gas. Active removal of carbon dioxide is therefore necessary in any enclosed life support system.

Total atmospheric pressure falls in an exponential manner on ascent to altitude, as the effect of the Earth's gravitational attraction declines and the influence of solar energy on gases in the upper atmosphere increases. There are parallel reductions in both air density and temperature. Figure 3.1 is an enlargement of the physiologically relevant part of Figure 1.1 and shows these relationships graphically. It is worth remembering that, since pressure is really a reflection of the number of molecular collisions occurring, there will be a point during ascent to altitude when such collisions become very rare. This point is generally taken to be at an altitude of 435 miles (700 km) and is the start of the so-called 'Space Vacuum'. The physiological limit for man occurs long before this, however, at an altitude of 60,000 feet (18,293 m): above this altitude, a spacesuit or cabin is essential for the protection of the human body.

Figure 3.2 shows the relationship between the partial pressure of oxygen in the lungs, the percentage saturation of the blood with oxygen, and altitude. The relationship is described by what is called a sigmoid curve with a plateau at the top. This plateau represents an in-built physiological reserve and is a function of the unique way in which oxygen combines with the haemoglobin contained within red blood cells. Healthy people normally have blood with characteristics which place them at the top right of the graph, where the alveolar partial pressure of oxygen is 103 mmHg (13.7 kPa): this corresponds with a figure of greater than 99 per cent saturation of the blood with oxygen, and of course with sea level. A fall in oxygen partial pressure to 60 mmHg (8.0 kPa) results in a fall in saturation of only a few per cent. Such a fall may occur at sea level in ill people, as described, but is also an inevitable consequence of a climb to an altitude of 10,000 feet (3,048 m). Further reductions in alveolar partial pressure then cause a precipitate fall in saturation of blood with oxygen and the development of hypoxia. As far as any air travel is concerned, including spaceflight, the implications of this are clear: ascent to altitudes greater than 10,000 feet is not possible without some form of protection. For those flying in civil aircraft, protection is accomplished by the cabin pressurization system which keeps the inside of the aircraft at a pressure equivalent to an altitude of less than 10,000 feet (and commonly of 6,000 to 8,000 feet (1,829 to 2,439 m) for extra safety). A similar concept is used in spacecraft, but then the situation is considerably more complicated!

Finally, before leaving the subject of oxygen for a while, it may appear obvious that the answer to all of these problems would be to equip spacecraft with an atmosphere of 100 per cent oxygen at sea level pressure. But such a solution, besides creating a terrible fire risk, has its own physiological hazards. Breathing an oxygen-rich atmosphere, particularly at sea level and during diving operations, may lead to symptoms and signs of oxygen toxicity in a relatively short time. Thus, an effect on the respiratory system will be evident if 60–100 per cent oxygen is breathed for longer than twelve hours. A sore throat, central chest pain, pulmonary congestion, and a cough are complicated after another twelve hours by severe lung damage with bronchopneumonia, fluid collection within the alveoli (pulmonary oedema), and lung collapse. Such changes occur even sooner if the victim is in a hyperbaric environment (i.e. diving). Effects on the central nervous system are also seen when oxygen is breathed under increased pressure and occur more rapidly than the pulmonary effects: at sea level pressures or less they are not the problem they are in space.

Decompression sickness

The reduction in total atmospheric pressure with altitude, illustrated by Figure 3.1 (see p. 41), may also lead to the development of aviators' decompression sickness: known colloquially as the 'bends'. This illness is similar to that seen in divers returning to the surface of the water after a deep dive, and its aetiology (cause) is believed to be the same: when pressure surrounding a fluid-filled body is reduced, gases within the fluid come out of solution, or evolve, in much the same way that bubbles develop and rise to the surface of a bottle of carbonated drink when the top is removed. If the gases so evolved are also relatively insoluble or inert, they remain in gaseous form and are free to travel around the body as bubbles. Furthermore, if the reduction in pressure causing bubble formation is of such a degree that the inert gas pressure comes to exceed the total external pressure, then a state of supersaturation exists and symptoms and signs of decompression sickness may then be expected. In humans and other animals, the inert gas involved is nitrogen.

Fortunately, the risk of decompression sickness is minimal if the ratio of the initial (starting) pressure of nitrogen in the atmosphere to the final total atmospheric pressure remains less than 1.5–1.8: a value termed the critical supersaturation ratio (CSR). This means that, for healthy individuals starting at sea level, this potentially very serious condition is virtually unknown at altitudes below 18,000 feet (5,488 m) where the CSR is just 1.6 (although bubbles have been located by ultrasound at as low as 9,000 feet (2,744 m)) and is extremely rare below 25,000 feet (7,622 m). Above 25,000 feet, however, where the CSR (the ratio of the pressure of nitrogen at sea level (600.4 mmHg) to the atmospheric pressure at 25,000 feet (282 mmHg)) is 2.13 and so exceeds the critical value, it becomes increasingly common and more severe the greater the altitude and the longer the duration of exposure. Once again, the passengers and crews of civil aircraft are protected by the pressure cabin but the altitudes to which space travellers may be exposed are clearly greater and their protection is correspondingly more complex. Spacecraft and spacesuit pressurization to a level which would not allow the CSR to be exceeded on decompression is one method of reducing the risks of decompression sickness, but this is not always technically possible or physiologically desirable. Another method involves the reduction in the concentration of inspired nitrogen by replacement with oxygen, even to the point of providing 100 per cent oxygen to breathe and eliminating all nitrogen from the body (denitrogenation by pre-oxygenation!). Oxygen is so rapidly utilized by the body that oxygen bubbles do not cause any problems even if they form because they are re-dissolved almost immediately. For manned spaceflight, a combination of these two methods is used, and protection against decompression sickness plays a vital part in the design and function not only of the life support systems for the spacecraft itself but also of any Portable Life Support Systems (PLSS) for use during activities outside the relative safety of the cabin; that is, during Extravehicular Activity (EVA).

The presence of nitrogen bubbles in body tissues may produce clinical manifestations according to the site of bubble accumulation. Thus, classical 'bends' is the pain felt in and around joints of the body: pain which characteristically gets worse on movement and has been likened to having glass or sand in the joint. Because aviators are normally sitting down, their arms are usually more active than their legs and so the large joints of the former are most commonly affected; but no joint is immune. Bubbles may also cause symptoms and signs in the skin (the 'creeps') with itching and rashes developing in the affected areas. Rather more sinister sites may be involved, however, and the small blood vessels (microcirculation) of the lungs may be flooded with bubbles which then cause difficulty in breathing, shortness of breath, and a painful cough (the 'chokes'). If the bubbles coalesce and manage to enter the systemic circulation (that is, the circulation other than to the lungs) the heart itself may be affected, as may the brain and central nervous system. In the former case, the heart would be unable to pump effectively, since air is compressible and so would not pass out of the cardiac chambers, with the result that the circulation would fail. Air bubbles around the brain and spinal cord may have an equally devastating effect and produce many various neurological symptoms and signs (collectively termed the 'staggers'), including profound shock and even death. Interestingly, however, recent evidence from ultra-sound studies has shown that bubbles are frequently present on ascent to altitude without producing any clinical manifestations and, although an individual pre-disposition has been recognized for many years, this important new work may eventually lead to a selection procedure for susceptibility to decompression sickness.

Ebullism

At an altitude of about 63,000 feet (19,207 m) the total atmospheric pressure has declined to 47 mmHg (6.3 kPa). This also happens to be the saturated water vapour pressure at a temperature of 37° C: that is, at body temperature. What this means is that, at that altitude and above, body tissues begin to vaporize or 'boil'. Small pockets of gas can be detected beneath the skin within two to three seconds after exposure to barometric pressures of 47 mmHg or less, and these rapidly coalesce to distend the subcutaneous tissues all over the body. Gas evolution can also be detected in the abdominal cavity after seven to ten seconds, and within the heart and great vessels after about twenty seconds. Although prolonged exposure to such events is not compatible with life, monkeys and dogs have successfully recovered from brief (up to two minutes) unprotected exposures to pressures that low. In cases where the animals have died, ebullism acts to complicate hypoxia which is the actual cause of death. Ebullism is obviously not a subject amenable to study in humans but subcutaneous distension (emphysema) has been seen in research workers and in some astronauts during training at low pressures without gloves. In these cases only the skin of the hand has been affected and, since mental function is retained, the subjects have been able to describe the phenomenon. Not surprisingly, there is a feeling of skin tightness, with pain and a prickling sensation, associated with an inability to perform delicate motor tasks. The condition resolved completely on recompression.

Barotrauma

Since the body may be regarded as being at a constant temperature, gases within its cavities are assumed to obey Boyle's Law. This states that, for a gas at constant temperature, volume is inversely proportional to pressure. Thus, on ascent to altitude, volume increases as pressure declines: at 18,000 feet (5,488 m) atmospheric pressure is half of that at sea level, and so any gas within the body will have doubled in volume. The lungs, the teeth, and the gut may be affected during ascent, while the middle ear cavities and the sinuses are usually affected during descent. Damage to tissues as a consequence of pressure changes is termed barotrauma.

Lung rupture may occur if there is a very rapid fall in external pressure relative to pressure within the chest, such as may happen during a rapid (explosive) decompression. During such an event, even an open airway and the wide bore of the trachea (windpipe) may be insufficient to allow venting of the rapidly expanding gas quickly enough to prevent excess pressure in the lungs. Rupture of lung tissue, with possible tearing of blood vessels and entry of air through the defects, can occur at over-pressures as low as 80–100 mmHg (10.7–13.3 kPa). Such catastrophic decompressions do not occur in the field of conventional aviation, but are clearly a very real possibility in space where the pressure differentials are so great.

Of much less serious consequence is the development of pain in a tooth on ascent; aerodontalgia. The precise cause of this phenomenon is not known but it is probably due either to irritation of the circulation in a diseased tooth or to a relative increase in pressure within a closed air space beneath a filling or a carious deposit. Since aerodontalgia does not occur in those with healthy or correctly restored teeth, this condition is just one reason why astronauts receive regular and strict dental examinations.

Expansion of gas within the small intestine may produce pain severe enough to cause fainting. This is only likely to be a problem in space flight if a rapid decompression occurs, when other effects will be considerably more life-threatening. Mild distension of the small bowel may occur during normal ascents to cabin altitudes greater than sea level, but even this may be minimized if a diet low in gas-producing foods (curries, beans, brassica vegetables, etc.) or fluids (alcohol, carbonated drinks) is followed. Gas in the large intestine and stomach does not cause problems since it can be released easily.

On ascent, expanding gas in the middle ear cavities vents easily via the Eustachian tubes: small canals connecting the cavities with the back of the throat. Only rarely is any discomfort felt. On descent, however, gas may be unable to pass back up the tubes and symptoms develop unless pressure between the middle ear cavities and the atmosphere can be equalized. A feeling of pressure on the outside of the eardrum rapidly develops into severe and progressive pain which is relieved only by a successful attempt to open the Eustachian tube, by swallowing, yawning, moving the jaw from side to side, or raising the pressure within the mouth by pinching the nose with the mouth shut while blowing out. If pressure cannot be equalized the eardrum will eventually rupture, with consequent relief of pain.

Similarly with sinus problems: expanding gas vents easily on ascent via tiny openings, or ostia, leading from the sinuses to the nose. On descent, increasing pressure can readily occlude these small openings and a pressure gradient develops. Severe and sudden knife-like pain is the result, possibly with an accompanying nose bleed.

Thermal injury

Figure 3.1 (see p. 41) illustrates the severe fall in temperature with altitude, but in space the laws of thermal physics are distorted, and the temperature to which man may be exposed will vary enormously depending upon the position of the Sun, metabolic activity, and the active or passive measures adopted to control spacecraft temperature. The environmental temperature extremes range from $-113°C$ to $+1100°C$ and, since man is able to survive neither a prolonged fall nor a prolonged rise in body core temperature of more than about $5°C$, protection against the temperature fluctuations of space is an essential and obvious requirement. In the event, the principal problem is one of excessive heat, generated by both men and equipment within the relatively small volume of a spacecraft. In the even smaller volume of a spacesuit, the matter of heat dissipation (dispersal) becomes crucial. These aspects are discussed fully in Chapter 6.

Ascent to altitude, then, carries with it profound physiological consequences. These consequences are as potentially harmful for those enjoying the convenience and comfort of modern air travel as for those engaged upon the exploration of space. By its very nature, however, the latter invokes a greater sense of wonder and urgency. Figure 3.3 summarizes the physiological problems associated with the physical changes in the atmosphere seen on ascent to altitude.

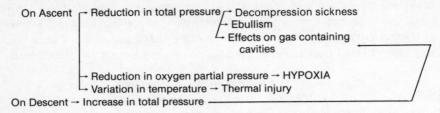

Figure 3.3 The physical effects of ascent to altitude and their physiological consequences

The solution – atmospheric control systems (ACS)

Quite clearly, the crews of spacecrafts must be protected against the hazards of reduced atmospheric pressure and density. The easiest and most obvious method of achieving such protection would be to provide, within the spacecraft, a cabin which mimics precisely the environment of Earth with respect to atmospheric composition and pressure; and, indeed, temperature and humidity. The occupants would then be breathing a 21 per cent oxygen 79 per cent nitrogen mixture at a pressure equivalent to that of sea level (760 mmHg (101.1 kPa)) appropriately warmed to a temperature of about 22°C and suitably humidified. This approach would not, however, provide protection against the hazards of decompression, which requires 100 per cent oxygen, and is not feasible for spacesuit (EVA) use, which requires a degree of mobility. For the reasons described on p. 43, it is also undesirable to provide an atmosphere composed entirely of oxygen at sea-level pressure. Furthermore, the requirements for space*craft* environments and those for space*suit* environments also differ in some vital respects, making the subject of atmospheric control in life support systems most complex. Compromises have to be made to accommodate these conflicting needs, and here the Russian and American approaches have been fundamentally different.

Spacecraft life support systems

A life support system (LSS) is any mechanical device that enables man to live, and usually to work, in an environment in which he could not otherwise survive. Such devices must allow the normal biological exchanges of gases, nutrients, waste, and energy between man and his surroundings to take place. Of these, the exchange of gases is the most immediately important, and atmospheric control systems (ACS) are the principal subject of this chapter. The problems of temperature and humidity, energy control, food, water, and waste management, with their implied additional constraint of time, will be discussed in Chapters 6 and 7.

The preceding section clearly established that oxygen must be supplied at adequate pressure and in sufficient quantity to maintain tissue oxygenation.

The problem of adequate pressure requires sophisticated engineering and several compromises to solve it, since different levels of pressure may be needed at different stages of spaceflight. The provision of sufficient quantities of oxygen, however, can be reasonably well predicted from the known metabolic cost (oxygen consumption) of various activities. A man at rest consumes about 250 millilitres (ml) of oxygen each minute. This figure approximately doubles when moderate exercise is undertaken, but may rise to 3 litres per minute ($L.min^{-1}$) when exercising at peak performance. By judicious analysis of the likely muscular effort required throughout a space mission, and by building in a safety element, sufficient breathing gas for a mission of a given length can be provided. It will be recalled that, in the Gemini programme, the planners 'got it wrong' when the metabolic cost of EVA was found to be much higher than predicted.

It is obvious that the oxygen supply for a spacecraft must be carried by it into space. The simplest way in which oxygen can be provided is from bottles in which the gas is stored under high pressure, but the weight and bulk of such cylinders preclude their use for all but short missions or as small emergency supplies. Liquid (cryogenic) oxygen, stored under pressure at low temperature in vacuum flask-like containers, reduces both the weight and bulk penalties, and prolongs the duration of supply (one litre of liquid oxygen vaporizes to form 840 litres of gaseous oxygen). Even lighter, and rather more stable, self-contained methods of oxygen production include the use of solid chemicals such as sodium chlorate. Candles of sodium chlorate, once ignited with iron filings, release pure oxygen as the chlorate decomposes. Other solid materials, the alkaline superoxides (sodium or potassium dioxide, or both), react with water or carbon dioxide to release oxygen from its chemically bound state. Finally, electrochemical reactions can also be used to derive oxygen from water or carbon dioxide. All of these methods, singly or in combination, have been used at various stages of the manned spaceflight programmes.

A further, attractive means of producing oxygen, the so-called molecular sieve, has not yet been incorporated into spacecraft technology because it requires considerable quantities of compressed air from which to concentrate oxygen. Such devices, currently being installed in modern fighter aircraft, consist of beds of zeolite material (hydrous silicates of calcium, magnesium, sodium, etc.) which are able to trap nitrogen molecules within their structure when air is delivered under pressure, but allow oxygen to pass through in large amounts. When the beds are depressurized, the nitrogen can be purged off the zeolite which can then be used again in the next cycle. There is no doubt that if an adequate source of compressed air was available on board spacecraft (for example, on board large space stations) a molecular sieve could be a cost-effective means of life support.

Molecular sieves have, however, found other uses in current spacecraft, including the removal or 'scrubbing' of carbon dioxide from the cabin atmosphere. In this case, too, the sieve can be re-cycled since carbon dioxide can be removed from the bed by heating. Another re-usable system is the amine scrubber, similar to that used in submarines, in which the carbon dioxide is absorbed by highly basic organic compounds (amines) to form carbonates. The material is then heated to reverse the reaction and to drive off the accumulated gas which is then dumped overboard. Other forms of carbon dioxide scrubbers are more common, however, and while they are simpler, they are not re-usable. This group of chemical scrubbers, including soda lime, Baralyme (barium hydroxide + calcium hydroxide), and lithium hydroxide, is supplied in refillable or disposable canisters through which cabin air is pumped. The carbon dioxide combines irreversibly with the chemical and uncontaminated air passes back into the cabin. The chemical reactions generate water and heat in varying amounts and the usefulness or otherwise of this, along with the economic factors of weight, bulk, and cost, tends to determine which chemical scrubber is used.

This, then, is the theory behind the oxygen supply to, and carbon dioxide removal from, a spacecraft environment. Cabin pressure is controlled by sealing the compartment at sea level prior to launch, and then delivering the appropriate gas mixture of oxygen and nitrogen to a predetermined pressure level. As the crew depletes the oxygen in the cabin, and the scrubbers remove carbon dioxide, the shortfall is made good by a continuous bleed of additional oxygen from the main supply. The concentrations of all gases and their pressures are therefore monitored and controlled continuously.

Russian systems

Despite the weight and bulk penalties, and engineering complexity, of a spacecraft pressurized to sea-level equivalent, the Russian space programme has utilized this system from its beginning. Thus, throughout the Vostok, Voskhod, Soyuz, Salyut, and Mir programmes, cosmonauts have enjoyed a sea-level environment, although they have been obliged to breathe pure oxygen at reduced pressure for long periods before embarking upon EVA, and indeed before take-off in order to reduce the risk of developing decompression sickness. Since the pulmonary effects of a 100 per cent oxygen environment at a sea-level equivalent pressure are toxic and so unacceptable, the option to provide an atmosphere of pure oxygen is not open to the Soviet scientists as a means of simplifying their chosen method of protection in space.

The cabins of Vostok, Voskhod, and Soyuz spacecraft were hermetically sealed at sea-level pressure before launch, and the system relied for its subsequent function upon being leak tight. Once sealed, a so-called closed loop regenerative ACS was employed in which cabin air was circulated continuously through a series of controlling units by means of an electric ventilating fan device. The most important of these units was the gas analyser which monitored the concentration of all constituent gases. Carbon dioxide levels were scrubbed to less than 0.5 per cent (i.e. 3.8 mmHg (0.5 kPa)) as cabin air passed through another unit containing superoxides: the regenerator. Oxygen was stored in its chemically bound state within these superoxides and released as required to replace oxygen consumed by the crew. The quantity released was controlled by the amount of water allowed into the regenerator, since it is the absorption of water which liberates oxygen; while the alkali so formed absorbs carbon dioxide. This elegant cycle was complicated by the fact that the flow through the regenerator needed to keep the carbon dioxide level low was such that too much water would have been delivered, so producing excessive amounts of oxygen! Consequently, the water content was also controlled, and reduced where necessary by drying with chemical absorbents (Vostok and Voskhod) or heat exchangers (Soyuz). Finally, if cabin pressure began to fall, a further regulating device increased flow through the regenerator. No additional supplies of atmospheric gases were carried. A display of all the measured variables was available to the cosmonauts and also relayed to the ground. For the six Vostok flights, a spacesuit was worn throughout and formed an integral part of the spacecraft LSS. The Voskhod and early Soyuz crews were able to dispense with their suits for much of the time but, following the rapid decompression which killed the crew of Soyuz 11, subsequent crews reverted to wearing spacesuits at all critical periods of flight. The atmospheric environment within Salyut and Mir space stations was also equivalent to that at sea level.

American Systems

In America, for the Mercury and Gemini flights, NASA chose a single gas system, with its advantages of simple engineering and savings in weight and bulk. An atmosphere of 100 per cent oxygen was provided at a pressure of 259 mmHg (34.5 kPa) (equivalent to an altitude of 27,000 ft (8,232 m)) which clearly minimized the risk of decompression sickness during the normal transition at launch from a sea-level pressure environment to the reduced pressure within the spacecraft. Protection was also provided, therefore, should inadvertent decompression occur. Furthermore, since oxygen toxicity is a function of the partial pressure of oxygen being breathed, these early American spacecraft environments were easily tolerated since an oxygen partial pressure of 259 mmHg is well below the toxic threshold.

Mercury The Mercury ACS was designed for flights of no more than twenty-eight hours' duration. It was a dual system, with one sub-system controlling the spacesuit environment and the other controlling the general cabin environment. As in the Vostok missions, the spacesuit was worn throughout flight and was fully integrated with the craft systems. Oxygen from a pair of high-pressure cylinders was fed into the spacesuit at the waist, circulated around the body and was then discharged from the helmet, sweeping across the inside of the visor to remove exhaled moisture before it did so, into a series of units which filtered debris, scrubbed carbon dioxide using lithium hydroxide, and modified temperature and humidity. The gas was then re-circulated through both the suit and the cabin. Oxygen consumption by the astronaut, and any loss through leaks, was replaced by additional oxygen from the bottles. If the spacecraft depressurized, the suit pressure was maintained at 238 mmHg (31.7 kPa). The dual system allowed independent control of spacesuit temperature which was especially important during the re-entry phase. After re-entry, the spacecraft was re-pressurized by atmospheric air during the final stages of descent and after landing.

Gemini In keeping with the aims of the project, including the support of two-man fourteen day missions and EVA, the Gemini ACS was improved in many respects and considerably more complex, although the 100 per cent oxygen atmosphere at 259 mmHg (34.5 kPa) was retained. Separate suit and cabin sub-systems were developed and the crew could choose to use either: indeed, during the long Gemini missions the astronauts were able to dispense with their spacesuits and to operate in their long underwear. The primary source of breathing gas to the suit sub-system was liquid oxygen but pressurized bottles supplied the cabin and were available for use in emergencies and during re-entry. As before, oxygen was fed into the suit at the waist and was then circulated around the torso, limbs, and head before leaving, this time not at the helmet but at the waist, to undergo filtering, scrubbing, cooling, pressurization, and replenishment. The system was modified slightly for use during EVA (see p. 56) and the capsule clearly had the ability to depressurize prior to such activity and to re-pressurize subsequently.

Apollo It was at first intended that the Apollo spacecraft would operate with a Gemini-style ACS but suspicions and concerns about the potential fire risks of using a 100 per cent oxygen atmosphere were tragically confirmed by the fatal fire on board Test Apollo 204 (Apollo 1) in 1967. For the manned Apollo flights the cabin atmosphere at launch was therefore changed to a 64 per cent oxygen 36 per cent nitrogen mixture at a pressure of, again, 259 mmHg (34.5 kPa). The partial pressure of inspired oxygen was therefore 165.8 mmHg (22.1 kPa); that is, just above sea-level equivalent. Pure oxygen was breathed for three hours prior to launch from the closed environment of a spacesuit so that the risk of decompression sickness was obviated during the transition from the Earth's pressure environment to the lower one of the spacecraft. It also provided protection against the possibility of an accidental decompression to even lower pressures during the launch sequence.

Once the spacecraft was in orbit, and the risk of decompression receded, the oxygen content of the cabin environment was gradually increased to greater than 90 per cent by purging the nitrogen through the valve normally used for discarding urine! The astronauts were able to remove the helmets and gloves from their main suits to improve mobility at certain stages of established flight; and at other times were able to remove the suit altogether (although the watch-keeping astronaut retained his suit at all times). During such periods, they wore so-called 'constant wear garments': essentially flame-proof underwear. If a decompression occurred during these times, the LSS was designed to deliver a flood flow of oxygen into the cabin in order to maintain pressurization at 181 mmHg (24.1 kPa) for the five to fifteen minutes it would take to re-don the spacesuits.

In any event, a sealed spacesuit was also worn during docking manoeuvres and re-entry; and, of course, during EVA and lunar activity. Oxygen was carried in a semi-gas, semi-liquid form in low temperature containers, and supplied three systems in basically the same way as those in Gemini had operated. The main Command Module (CM) system was supported by supplies from the Service Module until the re-entry sequence began and a second, self-contained, CM system was activated. A third system, designed to sustain two men, was installed in the Lunar Module (LM). Spacesuits were worn continuously when in the LM, although the gloves and helmet could be removed for short periods. The main oxygen supply was delivered first to the suits and then to the cabin, and also served to charge the suit Portable Life Support System before activities on the lunar surface were undertaken. Once again, lithium hydroxide canisters were used in all three systems to scrub the atmosphere of carbon dioxide. An optimal mission design limit for the partial pressure of carbon dioxide in the cabin was set at 3.8 mmHg (0.5 kPa) with a maximum limit for continuous exposure of 7.6 mmHg (1.0 kPa). In the event, the carbon dioxide levels were maintained near to 1.0 mmHg (0.13 kPa) for long periods during Apollo missions, although tensions approaching the emergency level of 15.0 mmHg (2.0 kPa) were encountered during the Apollo 13 flight.

Skylab The Skylab ACS incorporated yet more improvements, not least because its size allowed the carriage of heavier equipment. A two-gas, oxygen-nitrogen system at a pressure of 259 mmHg (34.5 kPa) was again employed, but the final oxygen concentration was reduced to 70 per cent (cf. more than 90 per cent in Apollo) to provide only a slightly higher partial pressure than back on Earth (181.3 mmHg cf. 159.6 mmHg (24.2 kPa cf. 21.3 kPa)). This ensured that any possible sequelae of chronic mild hyperoxia (too much oxygen) were avoided and that there was no interference with medical experimentation. The control system was automatic but could also be handled manually and indeed was during much of the experimental work. Oxygen was supplied from six liquid-oxygen tanks, while a two-bed molecular sieve was used to eliminate carbon dioxide. The latter operated at a carbon dioxide pressure of 5.0 mmHg (0.7 kPa) so that levels in Skylab were generally higher than those previously experienced, although still well below the maximum limit for continuous exposure.

Apollo-Soyuz Test Project The Apollo-Soyuz Test Project (ASTP) presented some unique problems for the system designers of both countries since the pressurization schedules of Apollo and Soyuz craft were entirely different in philosophy. Pressure within the Apollo was set at the usual level of 259 mmHg (34.5 kPa), but pressure within the Soyuz craft was allowed to fall from its usual 760 mmHg to about 517 mmHg (68.9 kPa). Repeated transfers were then made by both crews, to and from both craft, via the in-flight docking module which acted as an airlock. Pressure within the module was elevated or reduced according to the destination of its occupants. No problems of decompression sickness were reported. Indeed, decompression sickness has not been a major feature of either the Russian or American programmes. In his autobiography published some years after the events, however, Astronaut Collins did admit to two episodes of 'bends' pain in his left knee: once during his flight on Gemini 10, and again during the historic Apollo 11 mission when he was CM pilot. Neither occurrence was revealed to the mission controllers at the time.

Space Transportation System The Shuttle is the first American spacecraft to utilize a sea-level equivalent atmosphere, comprising 21 per cent oxygen and 79 per cent nitrogen at a pressure of 760 mmHg (101.3 kPa). This means that the occupants are at risk from decompression sickness should the cabin be depressurized accidentally for any reason during flight, and would be at a similar risk if no preventive measures were adopted before deliberate decompression prior to EVA. This problem will be discussed further in the following section. Oxygen is carried in liquid (cryogenic) form, as is the nitrogen required. The two gases are allowed to mix via an oxygen-nitrogen control valve which not only maintains sea-level equivalent pressure within the cabin, but also preferentially closes the nitrogen feed line, so allowing only oxygen to enter the atmosphere whenever sensors indicate that oxygen partial pressure has fallen below the desired level of 159.6 mmHg (21.3 kPa). Carbon dioxide is scrubbed from the circulating air by disposable lithium hydroxide canisters.

Portable life support systems

Portable life support systems (PLSS), of which spacesuits are vital components, must provide all the atmospheric needs of man discussed above. In addition, they must provide facilities for thermal and humidity control, some measure of waste management, protection against the hazards of direct exposure to the space vacuum (radiation and micrometeoroids), and both manoeuvrability and a means of manoeuvring. These aspects, which must also include a consideration of spacesuit construction, will be discussed in Chapters 5, 6, 7, and 8, although their separation from the ACS is somewhat artificial. Only the need for nutrition is not an essential requirement and can be confined to activities on board the mother ship.

Russian spacesuits

Russian spacesuits, which were of conventional multi-layered design and construction, have employed two main types of ACS. The first, used for the Vostok and Voskhod missions, was of an open-circuit design whereby gaseous oxygen was delivered to the suit for breathing purposes and to maintain pressurization. A through-flow of oxygen purged carbon dioxide and water vapour into the cabin where the atmosphere was subjected to the main ventilator-regenerator system described above. The same suit was worn during EVA, supplied with oxygen from tanks on the cosmonaut's back. A safety umbilical, which also carried other services, kept the space walker attached to his craft.

The second type of suit, used in later Soyuz missions, employed a closed-circuit, or regenerative, ACS. In this type of system, the semi-rigid suit was pressurized as before, to a level of 300 mmHg (40 kPa), with oxygen from a supply in the backpack but the expired gas was not vented to the outside. Instead, it was re-circulated through a series of units which removed carbon dioxide, adjusted pressure, heat and humidity, and replenished oxygen as required. Thus, it was a miniature version of the regenerative systems used for the spacecraft. An advanced lightweight suit, capable of operation at pressures as low as 176 mmHg (23.5 kPa), with consequent improvement in mobility, was used during recent Russian EVAs.

American spacesuits

Mercury The spacesuits worn by Mercury astronauts, like those worn by their Vostok and Voskhod peers, were really emergency systems designed to sustain life in the event of a cabin ACS failure or if abandonment at very high altitude became necessary. The suits were similar in design and construction to the full-pressure suits being worn at that time by pilots of high-altitude military jet aircraft, and their primary function was to act as restraining devices to maintain the internal pressure. A high degree of mobility was neither a requirement nor a possibility, because of the extremely cramped Mercury capsule; indeed, the garment formed an integral part of the spacecraft system.

Gemini The need to work outside the spacecraft during the Gemini missions led to considerable changes in spacesuit design, directed not only at enhancing protection against radiation and micrometeoroids but also at improving mobility while still retaining a pressurized suit, albeit to the slightly lower level of 191.3 mmHg (25.5 kPa) adopted during EVA. During such activities an oxygen supply was delivered from the main spacecraft system to the suit via a 7.6 m (25 ft) umbilical. The umbilical was covered with a thin layer of gold to aid heat dissipation, and it also carried other services such as electrical supplies, biomedical data lines, and radio-communication links. Oxygen entered the suit through a port in the neck seal, from a chest-mounted pressure and Ventilation Control Module (VCM), at a rate sufficient to flush through enough gas to maintain both carbon dioxide and temperature at acceptably low levels. The Gemini suit ACS was thus an open loop. In the event of an umbilical failure, there was a further ten to twenty minute supply of oxygen mounted in the suit itself.

An improved version of the chest-mounted unit, re-named the Extravehicular Life Support System (ELSS), was used for the second and subsequent Gemini EVAs. The ELSS pack was larger than the VCM and was intended to provide enhanced pressurization and ventilation performance, improved thermal control and an extended, thirty-minute emergency oxygen supply. In the event, both the VCM and the ELSS failed to cope with the metabolic load imposed upon them by the astronauts while working outside the spacecraft, and this aspect will be discussed further in Chapters 6 and 8.

Apollo Two types of spacesuit were used in the Apollo programme. The first was worn inside the Apollo cabin during launch, rendezvous, docking, and re-entry, and was of a straightforward multi-layered design. The second was designed to support activities on the lunar surface, often at great distances from the spacecraft. This EVA suit formed part of the Apollo Extravehicular Mobility Unit (EMU) and had even more layers than its intra-vehicular counterpart, although the ventilating and pressurization layers were the same. The other major component of the EMU was a truly self-contained PLSS carried in a backpack: there was no reassuring umbilical. The backpack weighed 66.0–86.2 kg (145–190 lb) on Earth (but only one sixth of this on the lunar surface because of the Moon's lower gravity) and provided oxygen for consumption and pressurization (again to 191.3 mmHg (25.5 kPa)), cooling water to the liquid cooling garment, power supplies, and communications. The oxygen was delivered to the suit and then returned to the backpack for purification and de-humidification before being re-circulated; that is, it was a regenerative system. Carbon dioxide was scrubbed by reaction with lithium hydroxide. Originally, sufficient oxygen for four hours' use was contained within the rechargeable backpack, but for the prolonged EVAs planned for Apollo 16 and 17 the PLSS duration was extended to seven hours. A further, emergency, supply of gaseous oxygen for breathing and cooling purposes was also carried. This entirely independent but non-rechargeable contingency supply, called the Oxygen Purge System (OPS), was designed to purge the whole system once manually activated, and to last for forty minutes: the time considered to be necessary to regain the safety of the LM. Happily, it was never put to the test. A remote control unit for the PLSS was mounted on the chest where the astronaut could control and monitor all the functions of the vital backpack.

Skylab For the few EVAs performed by the Skylab astronauts, spacesuits of the Apollo type were worn but with the services supplied not from a PLSS but from the spacecraft via an 18.3 m (60 ft) umbilical. Oxygen was delivered continuously and, unlike that of Apollo, the system was open loop.

Space Transportation System Since the astronauts on board Shuttle spacecraft enjoy an environment equivalent to that at sea level, decompression sickness is only a risk should accidental loss of pressurization occur, and there is no necessity for them to wear spacesuits at all during flight. Indeed, after the very sensible initial step of wearing suits for the Orbital Test Flights, and after the Challenger disaster, a shirt-sleeve environment for all crew members became the norm, even during take-off and landing; although a helmet incorporating an emergency oxygen facemask was worn during the launch and landing sequences as a precaution against toxic fumes or in case depressurization should occur. Future Shuttle crews will wear new lightweight partial pressure suits and parachutes during launch and re-entry. Exposure to reduced pressures during EVA also remains, since spacesuit technology has not yet developed an acceptably mobile suite for use at pressures above about 222.3 mmHg (29.6 kPa). Up to four hours of denitrogenation (i.e., breathing 100 per cent oxygen) was found to be necessary to ensure that the risk of developing decompression sickness was minimal on transfer from the Shuttle environment with 21 per cent oxygen at 760 mmHg to that of the spacesuit even filled with 100 per cent oxygen at 222.3 mmHg. Such a prolonged period of preparation before EVA was operationally unacceptable, and this dilemma is an excellent distillation of the conflicting physiological requirements of any ACS involving pressurization to sea-level equivalent. The logic of much of this chapter can now be summarized:

- The ideal spacecraft environment is one equivalent to that at sea level,
- but, such an environment will predispose to decompression sickness if accidental loss of pressurization occurs.
- This risk would be overcome if 100 per cent oxygen were breathed,
- but, breathing 100 per cent oxygen at sea-level equivalent pressure is a fire risk and is toxic to the lungs.
- Similarly, if the total cabin pressure was reduced from that at sea level while the oxygen partial pressure was held constant the risk of fire would increase,
- but, if the total cabin pressure was reduced and the oxygen partial

pressure was allowed to fall with it, hypoxia would become a problem.
- So, prolonged pre-breathing with 100 per cent oxygen is required before EVA is possible in a reduced-pressure suit,
- but such a requirement reduces the time available for EVA and is operationally unacceptable
- and, if spacesuit pressure were increased, mobility would be severely restricted.
- So, until a high-pressure suit with good mobility is developed, a compromise involving a period of intermediate decompression, and a consequently much shorter period of pre-breathing, must be adopted.

Although the three-and-a-half- to four-hour denitrogenation period remains one means of achieving EVA capability, and would be the only method available if EVA was required in an emergency, the Shuttle flight planners have devised an alternative method based on diving experience and involving an intermediate depressurization phase. Thus, before EVA is undertaken from the Shuttle, cabin pressure is reduced to 527.3 mmHg (70.3 kPa) – equivalent to an altitude of about 10,000 feet (3,048 m) – and the oxygen concentration increased to 27 per cent, that is, to an oxygen partial pressure of 142.4 mmHg (19.0 kPa). Although this *partial* pressure is equivalent to an altitude of about 3,900 feet (1,200 m) it clearly remains acceptable from the hypoxia standpoint. The whole crew then equilibrates at the new pressure for twelve hours before those members destined to leave the craft don their spacesuits and breathe 100 per cent oxygen for forty minutes prior to the final depressurization to 222.3 mmHg (29.6 kPa). (A similar technique was adopted by the crews of Salyut 7, which was also depressurized to an equivalent altitude of about 10,000 feet for some time before EVA was undertaken.) The Shuttle suit itself is part of a much upgraded version of the Apollo EMU with a total weight of about 110 kg (242 lb) on Earth. It incorporates the latest joint mobility technology and this allows it to operate at a pressure greater than that of its predecessors – 222.3 mmHg cf. 191.3 mmHg (29.6 kPa cf. 25.5 kPa). The PLSS backpack provides a suit environment of at least 95 per cent oxygen and has a duration of about seven hours, while carbon dioxide is scrubbed by a lithium hydroxide cartridge to a level of less than 7.75 mmHg (1.03 kPa) at expected metabolic rates. A thirty-minute emergency supply of oxygen is also carried. While in use, the PLSS oxygen and water supplies can be replenished from ports located in the Orbiter airlock.

Finally, a small Portable Oxygen System (POS) is provided for all Shuttle crew members, and can be used during launch, for denitrogenation prior to EVA, and during emergencies such as contamination of the cabin by toxic fumes. It is a closed (recycling) system and consists of a face mask, an oxygen regulator, and a lithium hydroxide canister to scrub carbon dioxide from the expirate. The regulator may be supplied either from the system's own small gaseous oxygen cylinder (independent or 'walk-around' mode) or from the Orbiter's main oxygen supply (dependent mode).

For future space missions, and particularly those on board craft utilizing 760 mmHg cabin environments, such as Mir and Space Station, a suit which operates at a pressure of 388–414 mmHg (52–55 kPa) is a highly attractive proposition, since there would no longer be any need to pre-breathe prior to EVA. Feasibility studies into surmounting the crucial mobility problems of such a suit are well advanced, and the next generation of EMUs will almost certainly be of this concept, although the ACS is likely to be similar in design and function to that currently in use.

Emergency life support systems

Once a spacecraft has been launched successfully, the most likely cause of a loss of pressure is penetration of the pressure cabin by a meteoroid. Because of this remote possibility, emergency systems have been designed to cope in terms of a specified maximum leak. Thus, the flood flow for Apollo, described above, refers to a puncture of 13 mm (0.5 in). Similarly, the Shuttle is designed to sustain a pressure of 414 mmHg (55 kPa) for 165 minutes (long enough for an emergency return to Earth) in the presence of an 11 mm (0.45 in) diameter hole. In this case, however, since 20 per cent oxygen in 414 mmHg is insufficient to prevent hypoxia, a supplementary oxygen supply is provided either from the POS or from the new partial pressure suit. Should the Shuttle be stranded for any reason and be unable to return to Earth, NASA has another, unique, emergency escape system under development. Each crew member will have a Personal Rescue Sphere (PRS), a fabric garment into which he or she can be zipped (Plate 5). The sphere can then be inflated with oxygen before being carried through space by a suited colleague to the rescue vehicle. Each sphere is 0.8 m (34 in) in diameter and has its own small supply of oxygen in the form of a POS, a window, and a small telephone! The PRS would also be used temporarily if the cabin air became contaminated. In such circumstances the cabin would be depressurized by two crew members wearing spacesuits, so venting the entire atmosphere including the toxic elements. For all other manned space programmes, however, the spacesuit has acted as the final refuge in the event of an emergency.

Finally, as an example of the flexibility of spacecraft design, the LM systems during the Apollo 13 emergency were able to sustain environmental control for nearly four days after the main supplies were lost.

Chapter four

Increased accelerations

The problems

Gravity is the force by which bodies on Earth are attracted towards its centre. In physical terms, gravity manifests itself as an acceleration (itself defined as the rate of change of velocity with time) and its magnitude expressed as distance per unit time per unit time. The normal gravitational force (g) experienced by bodies on Earth produces an acceleration of 32 feet per second per second (32 feet.sec^{-2}, 9.8 m.sec^{-2}) and for most people that force is more or less constant throughout their lives, and is experienced as weight. Small changes in the magnitude of this linear acceleration due to gravity occur every time a change in the speed or direction of motion of the body takes place: for example, accelerating during take-off in a Boeing 747 aircraft produces a forwards acceleration of just 0.25G. Some people, however, notably military pilots and astronauts are exposed to much greater accelerations as part of their normal working life. (Impacts as a result of car, train, or aircraft crashes clearly impose considerably greater stresses on individuals but these are not intentional and are often severely detrimental.)

The magnitude of these increased accelerations, and indeed decelerations, is conveniently expressed in multiples of the normal acceleration due to gravity, g, the figure then taking an upper case suffix, G. Thus, when a pilot says that he is in a 4G turn he is indicating that his aircraft is accelerating with a force four times that of the normal gravitational pull. Subjectively, he will feel four times heavier than normal. The direction in which the increased acceleration is acting is conventionally described by an axial diagram of the human body. From head to toe is the z axis, from chest to back the x axis, and from side to side the y axis. The axial diagram is shown in Figure 4.1. Accelerations are normally described as either positive (+) or negative (−), depending upon from which direction the force is applied, and that convention is used here. But the effects on the body of increased acceleration are in fact the consequence of the inertial force produced by that acceleration, and are thus always in the opposite direction to it in accordance with Newton's Third Law of Motion (to every action − accelerative force in this

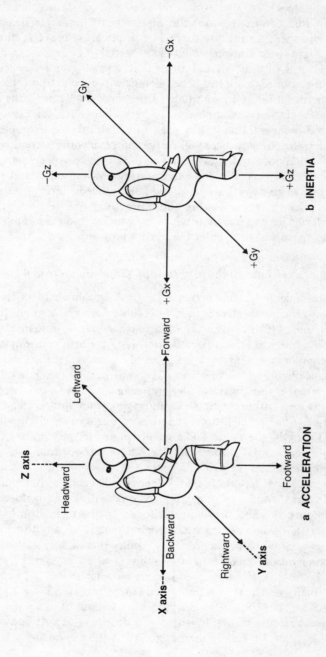

Figure 4.1 Axial diagram of human co-ordinate system for linear motion
a Direction of applied forces or accelerations
b Direction of consequent inertial forces (and of movement of body organs relative to the skeletal frame)

case – there is an equal and opposite reaction – inertial force in this case). Figure 4.1 shows the axial diagram of the directions of inertial (resistive) forces. (Strictly speaking, therefore, the terms +G or –G *acceleration* are misnomers in this context.) From the diagrams for example, headwards acceleration produces an inertial force in a footwards, or +Gz, direction. Similarly, forwards acceleration produces an inertial force in a backwards, or +Gx, direction. It is with the inertial forces that physiologists are concerned.

The human response to applied accelerations will depend upon the rate of onset, and the duration and magnitude of the increase. Thus, if the rate of onset is very abrupt and the applied force very large albeit for a very short period of time (conventionally, less than one second), for example in an aircraft impact, the consequences will be *pathological* with severe damage to body tissues. Such potential damage has implications for spaceflight during the final stages of return to Earth either during conventional land and sea recoveries or when ejection has been employed. Longer duration accelerations, that is, those lasting for more than one second, result in *physiological* changes and it is these which have influenced man's ability to be launched into space and to re-enter the Earth's atmosphere.

Long-duration accelerations: the problems of launch and re-entry

For most mechanically mobile activities, man is upright and usually seated facing forward. Linear accelerations, that is those acting in a single direction in a straight line, which last longer than one second, do not produce any important physiological effects because nothing on Earth can impose them for long enough, or at a sufficient level, to be of consequence: the same cannot be said of spaceflight, where velocity is less constraining (see p. 67). Radial accelerations, that is those achieved by imposing a change in the direction of movement, are seen within the Earth's atmosphere when fast military aircraft are manoeuvred. The accelerations experienced during such flight act in the long axis of the body, the z axis, and are colloquially termed 'pulling G' if the acceleration is headwards (positive Gz), inertial force footwards, and 'pushing G' if the opposite applies (negative Gz).

In conventional flight, positive Gz is experienced at the bottom of a dive or during inside turns; and it can be reproduced on the ground, for research and training purposes, in man-carrying centrifuges. In the relaxed, unprotected seated subject riding in a centrifuge, consciousness is usually lost by about +6Gz, although various other factors may modify this slightly such as rate of onset of acceleration, .height of the subject, and familiarity with the experience. If the rate of onset is sufficiently slow (less than 2G per second $(2G.\sec^{-1})$) then a clearly defined sequence of events will precede unconsciousness. At +2Gz, there is a subjective feeling of increased weight, with increased pressure on the buttocks and drooping of soft body tissues, especially of the face. By +2.5Gz there is great difficulty in raising oneself and by +3 to +4Gz it is impossible to do so. It is also difficult to raise the limbs. There is progressive dimming of the visual fields ('greyout') after three to four seconds with loss of peripheral vision (tunnelling) shortly thereafter at about +3.5Gz. Vision is then lost completely, a condition called 'blackout', although consciousness and mental function is retained. Convulsions occur either during or following unconsciousness in about half the subjects, with residual loss of orientation in time and space for about fifteen seconds. There is difficulty breathing and also congestion of the lower limbs with cramps and paraesthesiae (pins and needles).

The reason for all these features is the effect of acceleration upon the cardiovascular system: increasing acceleration produces changes in the hydrostatic pressure gradients throughout the body. Under the influence of the normal gravitational field, the pressure of blood throughout the arterial tree is uniform when the subject is lying down. Once standing, however, the column of blood between the heart and the brain exerts a hydrostatic pressure, creates a pressure gradient, and so reduces the effective pressure at the head. For example, if the hydrostatic pressure of the column of blood between the heart and the brain is taken at 25 mmHg (3.3 kPa) then an arterial pressure of 100 mmHg (13.3 kPa) at the heart will be reduced to 75 mmHg (10.0 kPa) at the brain. Since the pressure exerted by a column of fluid is the product of its height, its density, and the applied gravitational force, it follows that an increase in the last will further reduce effective pressure at the head. Unconsciousness may therefore be *predicted* to occur in the example described at +4Gz, where acceleration due to gravity is four times normal and the hydrostatic pressure consequently similarly increased. In fact, consciousness is usually retained even at +5Gz because several complex physiological mechanisms allow cerebral perfusion pressure to be maintained despite the fall in absolute arterial pressure. Firstly, the pressure *difference* across the walls of the intracranial vessels remains close to normal because hydrostatic pressures within the skull (acting as a rigid inexpansible box) fall in parallel with the vascular pressures at the head; secondly, resistance to flow through the cerebral circulation is reduced as the blood vessels dilate; and thirdly, the column of blood in the venous side of the circulation creates a siphon effect which essentially sucks blood through the brain to the heart for as long as the siphon holds. If the siphon breaks, as it will at higher levels of +Gz acceleration, consciousness is lost within three to four seconds.

Although the hydrostatic pressure changes above the heart have some effects, it is the changes occurring below that level which have even more significance. The increase in hydrostatic pressure below the heart leads to pooling of blood in the peripheral circulation: the blood vessels of the abdomen and lower limbs clearly neither have the benefit of being enclosed in a rigid container like the skull, nor are themselves rigid. The veins of these regions are particularly compliant and so dilate markedly under the influence of the increased pressure. Furthermore, such dilatation and filling leads to an increase in pressure across the walls of the blood vessels, and so fluid 'leaks' into the surrounding tissues. Thus, pooling of blood occurs in the lower body, less blood returns to the heart, and cardiac output is thereby impaired, so further compromising the circulation. Since this pooling is tantamount to a large haemorrhage, the body does invoke some protective mechanisms and, within five to ten seconds, venous return is augmented by constriction of peripheral blood vessels but it is never able to counteract the effects of acceleration completely. Clearly, similar but less marked hydrostatic changes are associated with the upright posture in a 1Gz environment. The cardiovascular physiology of man has adapted to these changes (see Chapter 11) but increased accelerations are able to outstrip the compensatory mechanisms.

The whole picture is further aggravated by an increasing failure of effective lung function. The pulmonary circulation is acutely susceptible to changes in hydrostatic pressure gradients, which means that, even at +1Gz (normal gravity), the tops of the lungs are not adequately perfused with blood, while the increase in lung 'weight' caused by +Gz acceleration tends to squash the bottom of the lungs so that ventilation with air becomes poor. The result of this ventilation-perfusion inequality is that the oxygenation of blood is diminished and cerebral hypoxia compounded. Even with protective manoeuvres, such as straining and muscle tensing, and protective devices, such as anti-g suits (which act by minimizing pooling in the legs) and reclined seats (which act by reducing the vertical distance between the heart and brain), the most highly trained and fit pilots will only be able to tolerate +10 to +12Gz for less than sixty seconds.

Negative Gz is even less tolerable for the normal human subject. Hanging upside down on a swing, an occupation undertaken by most children at some stage, is an experience of –1Gz, and it rapidly becomes extremely unpleasant as blood literally 'goes to the head'. During flight in conventional aircraft, negative Gz is experienced during outside turns or at the top of a climb. As might be expected, the physiological effects are the opposite to those experienced during +Gz acceleration. At –2 to –3Gz, there is pooling of blood in the head and neck, with unpleasant feelings of engorgement, a throbbing headache, and blurring of vision. Occasionally there may be a red misting of vision, so-called 'redout', as tears are stained with blood from conjunctival bleeding. Furthermore, the effect of raised hydrostatic pressures above the level of the heart is to simulate a reflex which slows the heart and may even cause unconsciousness. Levels of more than about –3Gz are painful, and are intolerable for any longer than a few seconds, so that manoeuvres which involve accelerations of this nature are avoided. The limit of human tolerance is about –5Gz and this is only rarely attained. Fortunately, the structure and strength of most modern conventional aircraft do not allow airframes to withstand negative G to any great extent either, so that both the man and his machine are constrained within suitable physiological limits. This does not apply to the forces encountered during the re-entry phase of spaceflight, where the increasing attraction of Earth's gravity, *if* it were to act in the long, z, axis of the body, would result in the imposition of considerably more negative G than is acceptable.

Short-duration accelerations: the problems of landing

Short-duration accelerations are experienced every day by those unfortunate enough to be involved in any impact accidents, such as those involving cars, trains, and aircraft. In these cases, the term acceleration is being used generically since such events imply sudden *de*celeration, the effects of which will depend largely upon the mechanical strength of the body tissues, since there is neither time for any physiological changes to become manifest nor any likely benefit to be gained from such changes. Assessment of human tolerance to impact is beset with difficulties since human subjects cannot be used experimentally in tests likely to produce irreversible damage, and dummies, while useful, are hardly fully representative. The medical and pathological evidence from accidents and from animal work is, therefore, extremely important. As with long-duration accelerations, the effects of short-duration insults will be determined by the site and direction of application, and especially by the time for which the acceleration acts: the shorter the period over which an acceleration is applied, the higher will be the peak level tolerated.

The matter is further complicated by the rate at which acceleration is applied. This parameter, known as jolt (G per second ($G.sec^{-1}$)), will produce greater injury when its value is high than when it is low, even though the peak acceleration may be the same. In some heroic American experiments during the build-up to manned space flight, men were catapaulted in a rocket-sled along a track into a water brake. This exception to the use of human subjects was an attempt to define the tolerance limits to impact and the most famous of these 'sled-riders' was an Air Force physician, Colonel John Stapp. In one experiment on the effects of jolt, he established that, in a forward facing impact, a peak of $-38.6Gx$ for 0.28 seconds produced no signs of shock when the jolt was 314 $G.sec^{-1}$, but that a peak of the same magnitude for 0.16 seconds when the jolt was $1315G.sec^{-1}$ produced severe shock. Put somewhat more simply, if Stapp and his colleagues were accelerated from a standing start to 632 $miles.h^{-1}$ (1,017 $km.h^{-1}$) in five seconds and stopped in less that 0.25 seconds, they experienced over $-40Gx$ and withstood a force of nearly five tons! This is equivalent to driving a motor car at 120 $miles.h^{-1}$ (193 $km.h^{-1}$) into a brick wall. The $-Gx$ axis was the one under test because, provided that full body restraint is applied, human tolerance to impacts is potentially greatest in this axis.

The rocket-sled experiments led to the conclusion that an impact of up to $-35Gx$ for 0.1 seconds would be tolerable without injury. Tolerance to impact in other axes is not so well-defined, largely because there are unknown regions between the areas of voluntary human tolerance and known injury levels. In the crucial +Gx axis (rearward facing), tolerance is critically dependent upon the adequacy of restraint and the posture of the body at the moment of impact. Volunteers representatively restrained have been uninjured by impacts of up to +32Gx when the impact velocity was 15.4 feet.sec^{-1} (4.7 m.sec^{-1}) and its duration just 0.001 second with a jolt of 1600G.sec^{-1}. Tolerance to lateral (sideways) accelerations (+/–Gy) is of the same order as that to +/–Gx accelerations provided that the neck is adequately restrained; otherwise, neck injuries occur during relatively minor impacts (+/–9Gy).

Impacts in the z axis of the body are uncommon in normal terrestrial life since we rarely travel vertically, but a falling lift would create a realistic scenario. In space, however, z axis impacts have been experienced and their effects, particularly on the integrity of the spinal column, are of more than theoretical interest. The word 'impacts' in this context is a slight misnomer since the concern is with the acceleration imparted by an ejection seat: the means by which military pilots escape from their stricken aircraft and the means by which the Vostok cosmonauts were routinely obliged to return to Earth.

Two main components comprise an ejection seat: the seat itself, in which the occupant is securely restrained, and the ejection gun. The gun is mounted in the z axis of the seat (and hence that of the pilot) and, on ignition, propels the combination away from the craft in that axis. The size of the explosive charges fitted to the gun determines the acceleration with which the seat departs and therefore the time required to clear the aircraft or spacecraft. It is quite possible to fit an immensely powerful gun and to eject a man within a fraction of a second but, in practice, the power is limited to that which will achieve a safe escape without damaging the spinal column. Application of +Gz accelerations greater than about 30G will almost invariably result in crush fractures of one or more vertebrae in the lower back. Furthermore, since there is often a dynamic overshoot which approaches this structural limit when the seat cushion and soft tissues of the buttocks, which are compressed initially, 'catch up' with the rigid structure of the rest of the seat, the initial thrust of an ejection is only about +18 to +20Gz and lasts for up to one second. But even this will produce crush fractures in many cases. In modern ejection seats the acceleration is sustained by the incorporation of further charges, usually in the form of seat-mounted rockets, which serve to carry the seat away from the vehicle and to help stabilize it in the windstream. Meeting the windstream exposes an ejected pilot to a deceleration in the –Gx axis, and there is a further deceleration of up to 20G as the main parachute opens. A parachute stabilizing system is then employed to further align the seat before man-seat separation occurs. The man descends to Earth beneath his parachute and lands at a velocity of about 21 feet.sec^{-1} (6.4 m.sec^{-1}), which is equivalent to jumping off a wall nearly seven feet (2.1 m) high.

The solutions

Launch and re-entry

As described in chapter 1, the physics of spaceflight mean that, in order to place a rocket into orbit 125 miles (201 km) above the Earth, a velocity of 17,900 miles.h^{-1} (28,800 km.h^{-1}, 8 km.sec^{-1}) must be achieved; while to leave the Earth's gravitational field, the velocity required is 25,950 miles.h^{-1} (41,753 km.h^{-1}, 11.6 km.sec^{-1}). These velocities may be attained by an infinite number of acceleration × time combinations, provided that the result is 828 G-seconds for Earth orbit and 1,152 G-seconds for Earth escape. From the previous discussion it is clearly impossible to impose an acceleration on a man of over one thousand G for one second, even if the rockets existed to achieve this. In fact, even the very powerful American Saturn and Russian launch vehicles can manage only a fraction of this thrust and only then for a relatively short period. It must also be remembered that re-entry to Earth's atmosphere and a subsequent safe landing requires considerable deceleration. Table 4.1 lists some of the salient particulars of vehicle launch (acceleration) and re-entry (deceleration) profiles.

Table 4.1 Some launch and re-entry acceleration profiles

Vehicle	Launch Profile	Re-entry Profile (average max G)
Vostok		8.0–10.0G
Mercury-Atlas	6.0G for 35 sec. and 6.4G for 54 sec. in two phases over 6 min. with peaks of 8.0G	8.9G (range 7.6–11.1G)
Voskhod		3.0–4.0G
Gemini-Atlas	Peaks of 5.5G and 7.2G	5.7G (range 4.3–7.7G)
Soyuz	3.5–4.0G	3.0–4.0G
Apollo-Saturn	Little>4.0G	5.9G (range 3.3–7.2G) for up to 60 sec.
Space Shuttle	3.4G	1.2G for 17 min.

In order that astronauts are able to tolerate these large accelerations, their bodies are orientated within the spacecraft so as to minimize the physiological effects of the induced hydrostatic pressure gradients: hence, during launch, the relevant human axis is +Gx (i.e., inertial forces from chest to back), with the crew lying on their backs. For re-entry, similarly (except in the Shuttle), the crews are fully reclined so that the inertial forces are from back to chest (i.e., –Gx). Human tolerance to long-duration Gx accelerations is primarily limited by effects on the chest and lungs. At +2 to +3Gx, there is some feeling of increased weight and abdominal pressure. Difficulty focusing and a slight feeling of spatial disorientation improve with experience and +2Gx is tolerable for at least twenty-four hours. In the range of imposed accelerations seen during spacecraft launches, +3 to +9Gx, there is increasingly severe chest pain, with difficulty breathing and speaking, as a result of the increased weight of the chest. Respiration becomes very shallow, with a reduction by half in the vital capacity at +6Gx. (The vital capacity is the maximum volume of air that can be moved in and out of the lungs: in healthy people it amounts to five or six litres.) Ventilation-perfusion inequalities will develop in the x axis, as they do in the z axis and, although they are less severe, arterial oxygen desaturation does occur. There is also a loss of peripheral vision, with blurring and lachrymation (flow of tears) making focusing difficult. Vice-like chest pain at +15Gx, with total loss of sensation and vision followed by unconsciousness, marks the limit to human tolerance of +Gx acceleration. On re-entry, −Gx acceleration produces similar effects but reversal of the force vector makes breathing easier. Chest pain and discomfort as a result of pressure against the restraint harness become severe at −8Gx, along with a feeling of insecurity. Visual distortion is also a feature.

It will be recalled that the angle of re-entry is crucial to its success. If the angle is too shallow the spacecraft will bounce off the atmosphere altogether, while if it is too steep the craft may not survive the extreme heat and the crew, even if reclined, will not survive the excessive deceleration. For example, the Apollo CM approached the Earth's atmosphere at a velocity of almost 25,000 miles.h^{-1} (40,225 km.h^{-1}) and was decelerated by the atmosphere to about the speed of sound (Mach 1: 770 miles.h^{-1} (1,239 km.h^{-1}) by the time it had reached an altitude of 30,000 feet (9,146 m). Further retardation was then achieved by the main parachutes, so that water entry was at the relatively sedate velocity of 22 miles.h^{-1} (35 km.h^{-1}).

Landing

From the above account, the potential impact forces on an unretarded space capsule would be immediately fatal: indeed, Cosmonaut Komarov died on impact when the retarding parachutes of his Soyuz 1 spacecraft failed to deploy after re-entry. Thus, retardation is a vital phase in the successful recovery of spacecraft. The parachute systems are designed to reduce the speed of descent, and the impact of the landing itself is further attenuated by techniques employed in the spacecraft construction. For example, the water-entry impact of Mercury spacecraft was reduced from about 50G to 15G by placing a fibreglass air cushion between the outer shell of the capsule and its heat ablation shield. The shield was designed to break away and, in so doing, to deploy the cushion. Project Gemini missions were originally intended to end with land landings, but development problems with the system meant that the back-up water-landing technique was eventually substituted. After re-entry, the capsule was retarded by the deployment of three parachutes. The first, a small preliminary drogue, deployed the second (pilot) and third (main) parachutes. These parachutes allowed the craft to reach an acceptable velocity before orientating it to the correct angle of attack for water entry (35° nose up).

Table 4.1 reveals a large range of re-entry accelerations for the Apollo missions but, in fact, the lower value represents that seen during Earth-orbit missions, while the higher values were seen during return from lunar orbit. This is not surprising because there was a considerable difference in the re-entry paths for the two types of mission. The Apollo landing system employed three main parachutes to lower the spacecraft into the water. These parachutes were deployed, closed, at an altitude of just below 30,000 feet (9,146 m) to be fully opened by about 10,000 feet (3,048 m). The most severe water impact was that experienced by Apollo 12 when the wind caused it to meet a wave at the wrong angle: the 15G landing displaced a camera from its mounting bracket in the cabin which hit an astronaut on the head, giving him a one-inch cut and rendering him unconscious for five minutes! With the advent of the Space Transportation System (STS) and its reusable Shuttle, parachute-retarded water-entry has become a thing of the past for the Americans. Landing in the Shuttle is a sedate procedure in the manner of a conventional aircraft, albeit without engines, with an exposure to just +1.2Gz for seventeen minutes prior to touchdown. During such a manoeuvre the seated posture would appear to be acceptable at first sight but the influence of even this level of acceleration may have detrimental effects on those with cardiovascular function adapted to microgravity (see Chapter 11).

With the exception of the Vostok missions, Russian spacecraft have also been retarded by parachute systems after re-entry: the landing is, however, on to land, the descent rate being further reduced by soft-landing rockets ignited automatically just before impact. Soyuz 23 demonstrated that the Russians could cope with a water landing when it landed in water, at night and in fog: all first-time events for the Russians and unintentional!

Finally, mention must be made of ejection seats since they were used routinely by the first six Russian cosmonauts, and were fitted as a means of emergency escape for the Gemini astronauts and for the crews of the first four (orbital test) flights of the Shuttle. The cosmonauts initiated ejection from their Vostok spacecraft after re-entry when at an altitude of 7,000 feet (2,100 m): they were still in their spacesuits but would no longer have required oxygen.

Emergency escape

All of the early spacecraft types were equipped with a means of escape in the event of an emergency, particularly during the launch sequence and the immediate post-take-off period. For the Vostok cosmonauts, the ejection seat provided this insurance. The ejection seats fitted to the Gemini spacecraft were intended for use not only in emergencies at the start of a mission but also as a primary landing system if the main parachutes failed or if re-entry over land was necessary. The seats could be used at altitudes of up to 60,000 feet (18,293 m) and so were provided with a self-contained means of protection against hypoxia.

Other spacecraft have employed a different approach: an escape rocket. The Mercury emergency escape rocket was designed to carry the capsule away from a stricken launch vehicle to an altitude from which a safe parachute recovery could be achieved. This meant that it was necessary for the astronaut to be exposed to considerably higher re-entry accelerations, up to +20Gx, than a conventional mission would impose. In this case, however, even if the astronaut was seriously injured, the alternative could not be contemplated. A similar escape mechanism was mounted on top of the Apollo-Saturn complex, but this time an aborted re-entry acceleration would peak at +16.2Gx. On the basis of centrifuge studies, it was concluded that, while crew members could probably tolerate such levels without serious injury, it was unlikely that they would be able to carry out any control tasks. An aborted re-entry was, therefore, almost fully automatic, with very little requirement for crew inputs. Soyuz spacecraft, too, have emergency escape rockets and their effectiveness was proved spectacularly in the successful abort during take-off of a Soyuz in 1983.

For the Shuttle, as indeed for the Apollo missions, in the event of a launch emergency while still on the ground, the crew access arm from the launch tower can be re-positioned within fifteen seconds. The crew would then be able to escape to the ground by the novel expedient of sliding down a wire in a steel basket. Five such slide wires run from the launch tower to the entrance of an underground bunker 375 m (1,230 ft) away, and each basket can carry two people. Shuttle crew members also wear a fabric harness during launch so that they can be easily manhandled out of the Shuttle during an emergency. After take-off, however, there are no provisions for emergency escape from a stricken Shuttle while it is travelling into orbit. This fact, combined with the devastating ferocity of the explosion which destroyed the Shuttle Challenger in January 1986, meant that the crew was doomed from the moment the fire started. For less catastrophic failures during launch and orbital insertion, several options exist which allow the Shuttle to return to Earth normally but prematurely.

Finally, emergency escape (egress) routes and drills are required for use after landing, for example from a sinking or burning spacecraft. Hatches have been the traditional method of providing such a facility, and that for the Apollo missions was extensively redesigned after the Apollo 1 fire tragedy. For the Shuttle, emergency escape after landing is normally via the crew entry hatch on the mid-deck but, if that is inaccessible, one of the overhead flight-deck windows can be used in conjunction with a rope and a thermal covering blanket to protect the crew from the high external skin temperature of the craft.

Chapter five

Radiation and micrometeoroids

Radiation

Electromagnetic radiation consists of a wave motion of electric and magnetic fields vibrating in phase at the same frequency at right angles to each other. Radiation is produced when there is a change in an existing electric or magnetic field. For example, when a charge is made to move up and down in a wire, such as a transmitting aerial, a changing electric field (current) is established which in turn sets up a magnetic field at right angles to it. This magnetic field then produces a further electric field which produces a second magnetic field, all again at right angles to each other. Repetition of the cycle produces a chain of electromagnetic waves which move away from the aerial at the speed of light. This is the basis of radio transmission, for if the transmitted waves meet a second, receiving, aerial they induce in it a current which matches the transmitting current. Since both the wavelength and the frequency of such motion can vary almost infinitely, a whole range of radiations can exist and this comprises the electromagnetic spectrum. (Thermal radiation is a specific form of electromagnetic emission caused by the agitation of a body's atoms and electrons.) At one end there are radio waves, with long wave lengths and low frequencies, and at the other there are gamma and cosmic rays, with short wavelengths and very high frequencies. In between lie infra-red rays, the visible spectrum, ultraviolet rays, X-rays, and rays produced by other atomic particles such as electrons, protons, and neutrons. Some of these, visible light and radio waves for example, are clearly beneficial; while others are capable of dissociating a substance in solution into its constituent ions and hence are extremely destructive to living matter. The former group comprise the non-ionizing radiations and the latter, with which this chapter is primarily concerned, the ionizing radiations (IR).

One of the most important functions that the atmosphere performs is to help protect the Earth's surface from the potentially devastating effects of radiated energy, of nearly all types, coming from space. There are, however, two 'windows' in the electromagnetic spectrum through which radiation from the Sun and from deep space may pass and so impinge upon the inhabitants of Earth. The first allows visible light and some of the ultraviolet and infra-red frequencies to pass; and the second allows high frequency radio signals to pass. All other radiation is either shielded by the atmospheric blanket or deflected by the Earth's magnetic field. Once beyond these protective layers, however, both space vehicles and their occupants are exposed to the full gamut of electromagnetic power, and particularly of IR.

Types of ionizing radiation

Various different types of cosmic IR, based largely on their source, have been identified by the many unmanned and manned space probes and missions launched over the past thirty-five years.

Galactic radiation   Galactic radiation originates outside the solar system, and is probably the consequence of major cosmic explosions. Deep-space probes have identified this radiation as consisting of protons (hydrogen nuclei) (87 per cent), alpha particles (helium nuclei) (12 per cent) and so-called heavy primaries (nuclei of heavy metals such as lithium, iron, and tin) (1 per cent). Some of these rays, and especially those produced by the heavy primaries, also termed HZE particles (High Z Energy where Z is the atomic number of the element concerned), may have components with extremely high energies. Since it is virtually impossible to shield passively against such energies, it is fortunate that the flow (or flux) of this type of radiation is very low. It has been postulated, however, that the flashes of light first reported by the crews of Apollo 11, 12 and 13 during times of darkness, and later confirmed by the Skylab astronauts, represented the impact of these heavy cosmic particles upon the retina of the eye after the rays had penetrated the spacecraft. Such penetration has been a relatively infrequent occurrence.

Solar radiation   Solar radiation, and particularly that from solar flares, is a major concern for manned space flight since the radiation produced, forming the solar wind, may be of both high energy and high flux (outflow). Solar flares are magnetic storms occurring on the surface of the Sun and their peak activity follows a ten- to eleven-year cycle. While the occurrence of a flare cannot be forecast, the build-up of activity can be monitored to provide an advance warning. The build-up of a flare takes place over several hours during which there is an increase in visible light and a disturbance in the Earth's atmosphere, probably as a result of solar X-rays. The flare itself can then last for days and produce a large flux of high-energy protons (10^6 to 5^8 electron volts).

Trapped radiation   Trapped radiation is the term given to the energy found in the two bands of geomagnetically trapped particles encircling the Earth. The magnetic properties of the Earth's substance have created a vast magnetic field about the planet which is able to hold or trap the electrons and protons of the solar wind as they approach. These particles then oscillate back and forth along the lines of magnetic force all around the Earth. The bands, or belts, were discovered by Dr James Van Allen in data sent back by the American Explorer satellites in 1958, and are named after him. The inner Van Allen belt starts at an altitude of 150–600 miles (241–965 km) depending on latitude, while the outer Van Allen belt begins at 4,950 miles (7,965 km) and may extend as high as 27,250 miles (43,845 km) depending upon solar activity. For all except the highest Earth orbital flights and flights to the Moon, the Van Allen belts have therefore presented no major problem (see Figure 1.1). Unfortunately, there is a discontinuity, or window, in the Earth's magnetic field in the southern hemisphere over the South Atlantic. In this area, known as the South Atlantic Anomaly, the effect of high energy protons can be detected at altitudes as low as 80–158 miles (129–254 km): well below the normal cruising altitudes of Salyut and Mir space stations, Skylab, and the Shuttle. Not surprisingly, when spacecraft are flying in this area, the flashes of light seen by the crew members increase in number.

Energetic neutrons Energetic neutrons have long been recognized as components of space radiation and as a potential source of great biological risk. This is because such neutrons are unstable and are liable to collide with hydrogen nuclei (i.e., with protons), of which there is a great abundance in the human body: proteins, fats, carbohydrates, and water all contain hydrogen. In colliding, there is a strong likelihood of an energy exchange and consequent damage to the tissues concerned. Measurements on board Skylab indicated that neutrons were present in larger than expected numbers, although fortunately not at a flux likely to cause damage, and that they were probably formed within the spacecraft as a result of trapped protons from the Van Allen belts bombarding its interior where they produced a secondary radiation. Their local production had been predicted because free neutrons are unstable, with a half-life of just eleven minutes, and decay into a proton and a neutron.

Although space radiation can take many varied forms it may be regarded as a single threat to the space traveller and is considered by many to be 'the primary source of hazard for orbital and interplanetary space flight'. Its potential danger for the future of long-term space missions will be discussed further in Part 3.

Units of measurement

An understanding of the effects of radiation on human structure and function requires some explanation of its measurement and expression. The *Roentgen* is the basic unit used to measure the quantity of radiation and is defined as the ionization produced by the passage of radiation (X-rays or gamma rays) through a medium (air) to produce a standard amount of air ions (0.001293 gram) carrying ± one electrostatic unit of electricity! This is a measure seldom used in the field of human biology. The *Rad* (the Radiation Absorbed Dose) is the most common measure of exposure since it takes account of the distribution of an absorbed dose and the effects of size, tissue composition, and changes brought about by internal interactions. One rad of any type of radiation corresponds to the absorption of 1×10^{-5} joules (the SI unit of energy) per gram of any medium: for living material the medium is tissue. A third unit, the *Rem* (Roentgen Equivalent, Man) describes the exposure of man to radiation most accurately. One rem represents that amount of any absorbed radiation which produces the same biological effect in man as would the absorption of one roentgen of X-rays. The rem measure is derived from the rad by applying a correction factor reflecting the relative biological effectiveness (RBE) of different types of radiation. The RBE is thus the ratio of a dose in rad of standard radiation (X-rays or gamma rays) to the dose in rad of the unknown radiation required to produce the same biological effect. Thus, X-rays and gamma rays have an RBE of one, while alpha particles may have RBE values as high as twenty. One rem may be regarded as approximately equal to one rad for the types of radiation encountered during manned spaceflight, particularly in low Earth orbit.

A fourth unit of measurement is required when dealing with the biological effects of heavy primaries in galactic radiation, since no other unit adequately describes their power. The *linear energy transfer* (LET) is the cosmic-ray energy (in kiloelectron volts) absorbed per micro-metre (1×10^{-6}m) of living tissue penetrated. That is, it is a measure of the density of ionization and will vary with the energy and type of IR, and with tissue composition. The higher the energy of a heavy particle, the higher will be its LET and its destructive power.

In 1985, in an attempt to clarify the confusing array of radiation nomenclature, the 'old' units of roentgen, rad, and rem were to be phased out in favour of SI units of radiation dose. These units have the dimensions of joules per kilogram and $1\mathrm{J.kg}^{-1}$ is equivalent to 100 rad; it has also been given the eponym Gray (Gy). Yet another named unit has been introduced to replace the unit of dose equivalence (rem) and is called the Sievert (Sv): and 1Sv is therefore equivalent to 100 rem! In this discussion, the old and familiar names have been retained.

Biological effects of ionizing radiation

In all living tissues, the effect of IR is to disrupt function at the cellular level: cell membranes and other components are deranged or destroyed, so interrupting or halting those functions basic to all cells and those specific to different cell types. The speed with which this damage occurs is remarkable: within 1×10^{-13} seconds cosmic particles enter the cell protoplasm and interact with each other to form ion pairs. These ions react primarily with cell water, within 1×10^{-9} seconds, to produce free radicals such as hydrogen or hydroxyl which then react together to form further active substances such as hydrogen peroxide. Even though these harmful substances are present for only a very short time (micro-seconds to seconds), and in very low concentration, they are able to devastate crucial cell components.

The biological effects of radiation upon man have been described extensively in the medical literature and have thus been dealt with primarily in terms of clinical studies of ill people. It is clearly unwise to extrapolate such information to healthy individuals without reservation, and this is particularly so for space travellers since many types of cosmic IR are not experienced on Earth and cannot be duplicated. Nevertheless, clinical experience does provide a good indication of the likely havoc which such radiation may wreak.

Whatever the effects, they will be influenced by and depend upon the types of radiation present and the duration and intensity of exposure. Furthermore, the anatomical site of exposure, the age and physical condition of the victim, and the complicating factors of increased and decreased accelerations and environmental control in space will all be relevant. Human tissues vary markedly in their sensitivity to radiation and are not affected equally. Those tissues which are actively growing, and so contain many dividing cells, are highly sensitive and are most at risk. This group includes, most importantly, the lymphoid system (comprising tissues which are responsible for the body's defence mechanisms), the haemopoietic system (responsible for the renewal of blood elements), the gastro-intestinal epithelium (responsible for the constant renewal of the gut lining), and the reproductive organs (the ovaries and testes). The next group, which demonstrates moderate sensitivity, includes the skin, lungs, kidneys, liver, and the linings of blood vessels. The lens of the eye is also in this group because it is particularly sensitive to neutrons; with opacities and cataracts being produced after quite low levels of exposure. Other tissues, including the central nervous system, bone, muscle, cartilage, and many glands, which reproduce slowly or not at all, form a third group with low sensitivity.

Not surprisingly, the clinical effects of exposure to radiation reflect the function of the most sensitive tissues and this can be seen in Table 5.1 which summarises the probable short-term effects of acute whole-body exposure to space radiation.

Based on this sort of clinical classification, it has been possible to define certain levels of radiation above which astronauts and cosmonauts should not be exposed. The limits are related to the likely total dose in rem considered to be a risk for certain occurrences. Thus, in the United States, a career limit of 400 rem to the bone marrow has been imposed on its astronauts by NASA, which is double the acute dose believed to induce leukaemia. Within this limit, other intermediate values have been set for a one-year average daily dose (0.2 rem to the bone marrow), a thirty-day maximum (25 rem), a quarterly

Table 5.1 Clinical effects of ionizing radiation

Dose (rem)	Clinical Effect
10–50	Minor blood changes.
50–100	Vomiting and nausea for one day; fatigue but no serious disability.
100–200	Vomiting and nausea for one day; 50 per cent reduction in circulating white blood cells within three days.

200–350	Prolonged vomiting and nausea, loss of appetite, diarrhoea and minor bleeding within four days;
	    marked platelet (clotting) dysfunction within six to nine days;
	    20 per cent dead in two to six weeks.

350–550	Same, but 50 per cent dead in thirty days
550–750	All have vomiting and nausea within four hours; nearly all dead within one week, but survivors ill for six months.
1,000	All have vomiting and nausea within one to two hours; all dead within days.

5,000	All incapacitated by nausea, vomiting and diarrhoea within minutes;
	    followed by disorientation, uncoordinated movements (ataxia),
	    shock and coma within minutes/hours;
	    no survivors.

maximum (34 rem), and a yearly maximum (75 rem). Similar reference risks have been listed for radiation received by other tissues such as the skin (career limit of 1,200 rem), the lens of the eye (600 rem), and the testes (200 rem). All of these figures refer to cumulative long-term exposures and not to acute episodes.

Radiation dosimeters

Radiation has been measured throughout most of the manned missions into space. Passive dosimeters were carried which consisted of sealed units containing various thermoluminescent chips (i.e., emitting light when hot), plastic sheets and metal foils each sensitive to a different type and energy of radiation. These devices were used for precise data-gathering and their results analysed only after return to Earth. Other so-called active pocket dosimeters were also carried and could be read at any time to provide the crew with a means of monitoring the radiation risk. Three types were used during the American space programmes, each capable of measuring a different range of radiation: low (0–200 millirad (1×10^{-3} rad)), high (0–100 rad) and contingency high rate (0–600 rad). At no time so far has either an American or a Russian crew been exposed to unacceptable levels of radiation. For example, the mean skin dose each day for an astronaut in the Gemini programme was 0.05 rad, while the mean cumulative dose was only 0.13 rad (range 0.01–0.84 rad). For the Apollo missions, some of which included traverses of the Van Allen belts, the mean daily skin dose was 0.044 rad, and the mean cumulative dose 0.41 rad (range 0.16–1.14 rad). The long duration

Skylab 2 (twenty-eight days), 3 (fifty-nine days) and 4 (eighty-four days) missions resulted in the relatively higher, but still safe mean cumulative doses of 1.98, 4.71, and 7.81 rad respectively. An eighty-four-day mission could therefore be flown each year for 150 years before a crew member exceeded the NASA career limitation for skin! The Russian experience has been remarkably similar, with a mean daily skin dose of 0.02 rad and 0.018 rad during the Vostok/Voskhod series and the early Soyuz missions respectively.

These data, however, are derived from relatively short-duration space missions and some projections have suggested that a crew member on Space Station will receive about 15 rad per ninety-day mission: this is higher than the acceptable equivalent exposure rate for any Earth-bound occupation and may well produce a detrimental effect after ten such missions. It has been claimed that this total dose of 150 rad will reduce the life expectancy of its recipient by an equal number of days, as a consequence of the increased risk of cancers. But this figure can be put into some perspective when it is realized that the length of a coal-miner's life may be expected to be reduced, by virtue of the risks associated with that occupation, by some 1,100 days; while even that of those engaged in seemingly harmless occupations, such as teaching, is reduced by thirty days. Furthermore, the levels of radiation to which astronauts have actually been exposed so far compare very favourably with those experienced from sources on Earth, such as routine chest X-rays (which deliver 0.01 rad to the bone marrow) and other medical procedures (such as barium studies of the digestive tract which can deliver up to 0.6 rad), during everyday life. Just living in Houston, Texas (the home of NASA's Manned Spaceflight Centre) for one year exposes the bone marrow to 0.1 rad!

Protection against ionizing radiation

Thus far, *passive* physical methods of protecting spacecraft crews have proved to be successful: that is, the radiation has been attenuated by the thickness of spacecraft walls and by arranging materials and equipment within the capsule so as to provide additional shielding. Shielding is normally expressed as units of density in grams per square centimetre ($g.cm^{-2}$), and a value of $2.0\,g.cm^{-2}$ has been calculated to provide sufficient protection against all but the largest solar flares. It would also keep the annual dose below 100 rad. The aluminium and stainless steel walls of the Apollo CM actually had protective values of $2.75\,g.cm^{-2}$ to $212\,g.cm^{-2}$, the latter being placed over the vulnerable base of the capsule. Other spacecraft have been similarly well-protected, although the Apollo LM had walls with a density of only $1.5\,g.cm^{-2}$: insufficient to protect against intense solar activity.

The low levels of radiation so far encountered, however, must reflect the careful planning of flight paths to avoid areas such as the South Atlantic Anomaly, and the fortuitous avoidance of times of unpredicted increases in solar activity. As long-term space stations become a reality, other methods of radiation protection may become necessary since any increase in wall thickness would carry considerable penalties of weight and bulk. As an example, a spacecraft should ideally have a protective shield equivalent to that provided for the Earth by the atmosphere but, in order to provide this equivalent, it would have to be surrounded by a water barrier 10 m (33 ft) thick or a solid lead shield 1 m (3.3 ft) thick. Moreover, as thickness is increased, a limit is reached beyond which the effects of secondary radiation induced by galactic cosmic flux outweigh the potential protective benefit. The use of internal equipment and stores also has limited applicability and, of course, some on-board equipment may produce IR in its own right and so require additional shielding. Other methods of protection must therefore be sought, although the passive local shielding of particularly vulnerable areas of the body, such as the abdomen, would undoubtedly reduce the severity of radiation sickness and improve its mortality.

The technology exists, although it has yet to be developed, whereby *active* physical protection can be provided by inducing an electric or magnetic field around a spacecraft. One such method involves the production of a current in a magnetic field by super-cooling super-conducting materials which then have zero electrical resistance and can provide circulation of a current almost indefinitely. This method has the advantages of being of low weight and of having low energy requirements. It can also be adjusted to cope with periods of increased radiation.

Finally, experiments in animals have identified many chemical substances which, after administration, have been shown to have a beneficial effect on the response to subsequent irradiation. The high doses and potential side effects preclude the use of this approach in humans at the moment.

Micrometeoroids

Meteors are small bodies of solid matter moving randomly through space. They are rendered bright and visible by the heating effect of compression or friction of air about them if and when they enter the Earth's atmosphere, and the largest are seen as 'shooting stars'. Any remnants which reach the surface of the Earth are termed meteorites or micrometeorites: most of the extraterrestrial material which strikes is in the form of the latter and, astonishingly, nearly 10 million kg (10,000 tonnes) of such interplanetary dust are estimated to reach the Earth's surface each day. While these objects are still in space, however, they take the names meteoroids and micrometeoroids respectively and form a collision risk for the space traveller. The magnitude of the risk has never been fully established and so it has always been the policy of the manned space flight programmers to provide passive shielding against micrometeoroid impacts. Most of these objects are made of stone (61 per cent), while the rest are either iron (35 per cent) or of mixed composition (4 per cent). Although they are usually small, they may have considerable energy and may form shallow craters in metal sheets. The spacecraft walls are thick enough to provide protection for most of the time spent in space, but for EVA and lunar activity a special lightweight multi-layer garment was incorporated into the spacesuit. The layers of this garment, including rubber-coated nylon, dacron, and teflon-coated Beta cloth, have proved entirely effective for prolonged periods outside the spacecraft. Experiments on board Skylab, however, did reveal that impacts were occurring over the test area ($1200\,\mathrm{cm}^2$ ($186\,\mathrm{in}^2$)) at the rate of nearly two per day. The largest impact measured was from a micrometeoroid believed to be about 10–20 cm (4–8 in) in diameter, but even this did not compromise the 3.18 mm (0.12 in) thick wall of Skylab despite the loss of its primary micrometeoroid shield during launch. Although the potential risk of micrometeoroids in space has not materialized so far, it is entirely sensible that passive precautions continue to be invoked; especially since the hazard has been found to be so slight that protection is easily accomplished.

Chapter six

Temperature and humidity

With due regard to the problems dealt with in the previous chapters, it would be perfectly possible for a cosmonaut or an astronaut to travel into space and return safely. Indeed, the crews of the very first Vostok and Mercury spacecraft did precisely that: they were provided with sufficient oxygen at adequate pressure, with suitable protection against the accelerations of launch, re-entry, and landing, and with shielding against the dangers of ionizing radiation and micrometeoroids. There are, however, several other vitally important aspects of spacecraft technology which must be considered if any manned missions longer than a few minutes or hours are to be undertaken. These aspects include the problems of thermal control, the control of humidity, provision of food and water, the disposal of human waste products, and other requirements for personal hygiene. Obviously, these elements of environmental control and life support systems were addressed for all the manned spaceflight programmes, including Vostok and Mercury, but were clearly of increasing importance as mission durations were progressively extended. This chapter and the next deal with these additional facets of spacecraft environments.

Thermal Control

Problems of the heat

From the brief descriptions in Chapters 1 and 3 it will be recalled that, although the range of temperatures to which man may be exposed in space is extremely wide ($<-113°C$ to $>+1200°C$), the main concern within spacecraft and spacesuits is the dissipation of internally generated *heat* and the maintenance of thermal comfort. Thus, while the Sun is very hot and deep space very cold, it is from within that the greatest risk of thermal injury lies.

Man is a homeotherm which means that, as in all higher animals, deep-body or core temperature is maintained within a narrow range and is largely independent of the temperature of the environment. Core temperature is that measured deep within the abdominal cavity and is maintained at a level of

37° C, with a diurnal variation of +/–0.5° C: maximum body temperature is achieved at about 5.00 p.m. for most people, while 3.00 a.m. is the usual time of day for the minimum temperature. The skin, the limbs, and the head and neck, which are obviously in more intimate contact with the environment, are often at a significantly lower temperature, even under normal conditions. For example, in a room at 20° C, exposed extremities may be at a temperature of just 28° C because body heat is lost to the cooler surroundings. A steady state therefore has to exist between heat production and heat loss in the normal individual so that a constant core temperature can be maintained. The body's metabolic processes produce heat continuously. At rest, over 50 per cent of total heat production is generated by the abdominal organs, while skin and muscle contribute a further 20 per cent; during exercise, however, heat production increases proportionately and working muscle may contribute up to 90 per cent of the total.

Whatever its origin, deep-body heat is transferred to the blood and carried to the skin where any excess is then lost from the body's surface by the processes of radiation, conduction, convection, and evaporation. Radiation is here defined as the transfer of heat from one body to another. The magnitude of its effect is related to the temperature of the radiating body and is unaffected by the temperature of the intervening air: hence, the Sun is able to warm the Earth and all upon it. Conduction is the exchange of heat between two bodies in contact with each other but at different temperatures. The magnitude of this effect depends upon the gradient, or temperature difference, between the two, and heat can be transferred in this way from the human body to air. Convection is the means by which dissipation of conducted heat can be enhanced by mass movement of air (or fluid) over the body. Such movement creates and maintains the thermal gradient necessary for effective conduction. Thus, the cooling effect of a breeze or a fan is achieved by replacing the warm and moist air around the body with cooler and dryer air.

The efficiency of heat loss by radiation, conduction, and convection can be further improved by evaporation of sweat; indeed, some heat loss by this method is essential during heavy exertion and, at environmental temperatures above 36° C, evaporation is the *only* effective means of losing heat. At environmental temperatures above 37° C heat is transferred *to* the body by radiation and conduction: perspiration then becomes profuse in an attempt to lose heat and so offset uptake from the surroundings. If such surroundings are also humid, as they are in tropical jungles, water cannot evaporate from the body's surface and environmental temperatures of only 34° C are intolerable. Provided that lost salt and water is replaced, however, much higher temperatures in dry heat, as in a desert, can be tolerated because sweat can readily evaporate.

Despite the considerable variations in heat uptake, generation, and loss, thermal comfort is maintained as far as possible by very powerful physiological reflexes and equally powerful behavioural responses. Thus, when hot, there is a physiological reflex manifest as an increase in blood circulation to the skin to enhance heat dissipation. This is especially true of circulation to the hands, feet, and face where sweating occurs, and to the limbs where surface area is large: radiated and conducted heat loss is then augmented by evaporative and convective loss. Depletion of body water and salt also stimulates the thirst centres in the brain and fluid intake is increased. Behaviourally, clothes are shed, shelter from the heat source is sought, and artificial means of cooling, such as fans, are invoked.

If the heat load is sustained and the physiological response is prolonged, the loss of fluid as sweat may lead to water and electrolyte imbalance with dehydration and other features of clinical heat stress or exhaustion. The victim will be agitated, restless, and distressed, and may suffer from headaches, nausea, and muscular cramps, particularly in the lower limbs and abdomen. The skin will be pale, clammy, and cold; indeed, body temperature remains normal and may even fall. Both respiration and heart rate will be rapid, with the former being shallow and the latter weak. If exercise is then undertaken and deep-body temperature continues to rise, even if fluids and salts are replaced, the risk of developing heat-stroke is considerable since the heart is unable to meet the circulatory demands of both the exercise and of the increased blood flow to the skin. The symptoms and signs of heat exhaustion are superseded in heat-stroke by a clinical picture of a victim with a full and bounding pulse, noisy respiration and very hot, flushed, dry skin (up to 40°C). Deep unconsciousness may develop suddenly and the condition will be fatal if not treated promptly.

Problems of the cold

When cold, we respond physiologically by decreasing peripheral circulation, so preventing heat loss from the skin, and by shivering to produce heat locally within the skeletal muscles. Behaviourally, voluntary muscle activity is undertaken, extra clothes are put on, shelter is sought, and external sources of warmth, such as fires, are established. Fortunately, medical problems of cold have not been a major feature of any manned space flight so far, but it is entirely possible that cold injuries could be seen in the occupants of a spacecraft whose life support system has failed, or if recovery to Earth has been complicated by delayed rescue from a cold and remote landing site on land or in the sea. Local cold injury, or frostbite, occurs when tissues of the extremities (fingers, toes, ears, nose, and chin) literally freeze on direct exposure to intense cold. The affected area is usually painful and the skin is hard, white, and stiff. Gangrene may develop in severe cases. Of far more sinister consequence is the generalized fall in deep-body temperature which results in hypothermia. The condition can develop when core temperature has fallen to just 35°C and will usually be irreversible if the temperature reaches

25°C. As body temperature falls over this range, subjective feelings of being miserably cold are rapidly succeeded by objective signs: the victim looks pale and feels abnormally cold to the touch, intense and uncontrollable shivering develops, and a state of lethargy may be noticed. Later, the shivering is replaced by muscular incoordination, irrational behaviour may be evident, and both respiration and heart rate decrease to become almost un-detectable as the victim loses consciousness. Death may follow rapidly, especially if the cause of the hypothermia is immersion in cold water.

Because of these very real risks, some consideration has been given to the survival aspects of landing from space in a hostile environment, be it cold and wet (sea), cold and dry (Arctic), or hot (desert). Russian and American space crews undergo survival training as part of their normal preparation for flight, and their spacecraft are equipped with survival aids including protective clothing, flotation garments, and emergency supplies of food and water. In April 1975, such training and equipment proved invaluable to the crew of a Soyuz (designated Soyuz 18A by some Western observers) which underwent an emergency re-entry from an altitude of about 90 miles (145 km) after a rocket separation failure during launch. The spacecraft landed in the mountains of Siberia just 200 miles (322 km) from China and the cosmonauts were obliged to 'survive' for over twenty-four hours before being rescued!

The physiological mechanisms described, however, normally allow the maintenance of a constant body temperature despite environmental temperature variations over the range 0°C to 50°C. The addition of behavioural responses extends this ability into areas, such as space, where survival would not otherwise be possible. Extremes of temperature obviously have clinically deleterious effects on the human body but the effects of a relatively minor alteration in the thermal environment may also have important operational consequences. The discomfort felt when either too hot or too cold, even though of a relatively small quantitative nature, will degrade mental and physical performance long before clinical manifestations are evident. The maintenance of thermal comfort is therefore of extreme importance if space crews are to perform optimally.

Environmental Control Systems (ECS)

It has been the intention throughout the manned spaceflight programmes to maintain a thermally acceptable or neutral environment about the cosmonauts and astronauts at all times. For unclothed men, the ideal thermal environment is one at about 30°C but this cannot be the case in space, where protective garments are essential. The temperature inside spacecraft cabins has therefore been controlled within the range 11°C to 27°C, which is a comfortable level for moderately clothed humans.

Temperature control is closely related to the control of other environmental factors such as atmospheric composition and humidity: the whole comprising the Environmental Control System (ECS). (Chapter 3 dealt with the atmospheric control sub-system of the ECS.) The dissipation of excess heat, both from men and equipment, is accomplished by the cooling sub-system of the ECS and several different techniques have been utilized.

Cabin temperatures in Vostok, Voskhod, and Soyuz spacecraft were controlled within the range 11°C–25°C by means of liquid-air heat exchangers. In these systems, cooling agent was passed continuously through an internal wall-mounted radiator, the exposed surface area of which could be varied by controlling its cover. A thermal sensing device automatically retracted the cover as cabin air temperature rose and closed it as temperature fell.

In the Mercury programme, environmental cooling was achieved by a sublimator heat exchanger: water vapour was produced as warm air was delivered to a tank of cold water. (Strictly, sublimation means the conversion of a substance in its solid state, in this case water, to its vapour form by the application of heat.) The water vapour was then vented overboard and the change of physical state in the system produced cooling. Additional water was supplied to the sublimator from the water separator used by the main ECS. A similar sublimating system was used as a secondary means of cooling in the Gemini spacecraft. The primary mechanism of heat loss, however, was via a spacecraft wall radiator, located in the Adaptor Module, which radiated heat to the cold vacuum of space. Control was achieved by varying the flow of coolant through the radiator and through the heat exchangers in the ECS. Space radiators were again used for heat dissipation in the Apollo programme but their efficiency was improved by slowly rotating the Command and Service modules so that all four radiators were presented equally and regularly to the cold of deep space: so-called Passive Thermal Control (PTC). (This rotation, in combination with the peculiar chequer-board black-and-white paint scheme used on the outside of the Saturn launch vehicles, allowed control of temperature by slowly presenting different surfaces of the rocket, with different thermal emissivities because of the paint scheme, first to deep space to lose heat and then to the Sun to receive heat.) When effective radiator operation was not possible, during the launch, Earth and Moon orbit, and re-entry phases, heat rejection was augmented by the use of glycol evaporators.

Cabin temperatures were successfully maintained at 24°C +/-2.8°C throughout the Apollo programme, with the exception of the emergency on board Apollo 13 when depressed electrical power levels led to uncomfortably low cabin temperatures of only 10–13°C. For the Skylab Orbital Workshop, the varying thermal emissivity of the external paint scheme was again exploited and very little active control, either by radiators or by evaporators, was needed although such devices were carried. Thermal control in the Shuttle is once more achieved by the use of curved space radiators mounted on the inside of the cargo bay doors so that they are exposed when the doors are opened in orbit.

Metabolic cost of work

Heat production from metabolic processes continues even when the body is completely at rest: the usual basal metabolic rate is about 35–40 kilocalories of heat produced per square metre of body surface area per hour $(kcal.m^{-2}.h^{-1})$, and is a direct reflection of the volume of oxygen consumed during that time. Any physical activity will increase the metabolic (oxygen) requirements of the body and the metabolic rate will rise correspondingly. Many studies have been undertaken to establish the metabolic cost of most types of human activity. For example, the metabolic cost of washing and dressing is about $87\,kcal.m^{-2}.h^{-1}$, that of walking is about $145\,kcal.m^{-2}.h^{-1}$, and that of bricklaying may be as high as $170\,kcal.m^{-2}.h^{-1}$. Sporting activities, particularly endurance events, are even more expensive metabolically and Olympic oarsmen may produce over $900\,kcal.m^{-2}.h^{-1}$, albeit for short periods of time. Not surprisingly, sedentary occupations tend to be relatively inexpensive since the seated posture relieves the body of a large energy demand. Driving a car produces a metabolic rate of about $93\,kcal.m^{-2}.h^{-1}$, while routine flying in a military fighter aircraft is even less expensive at about $83\,kcal.m^{-2}.h^{-1}$. The energy cost of work performed by individuals within the confines of a spacecraft cabin has also been shown to be low: the average metabolic rate of cosmonauts and astronauts during routine flight in early Vostok and Gemini spacecraft being about $50\,kcal.m^{-2}.h^{-1}$. But no matter what the activity, any extra heat produced must be dissipated if the problems described earlier are to be avoided.

In the spaciousness of an Earth-bound environment, such heat is usually easily dispersed; and even in spacecraft cabins, thermal control and comfort have been successfully achieved. The situation is completely different, however, once activity outside the spacecraft is considered and the individual begins to work within the very confined volume of his protective spacesuit. Large heat loads may be generated, not only by metabolic processes but also by several spacesuit facilities, including electric pumps and fans, and in some cases by the chemical absorption of carbon dioxide. In addition, attention must be paid to protection against the environmental temperature extremes which may be encountered. Although all of these aspects had been carefully considered prior to the first American space walks during the Gemini programme, the metabolic cost of EVA had been seriously underestimated and the ability of the spacesuit to deal with the heat load was often exceeded. Metabolic rates were not measured directly during the Gemini EVAs but several astronauts were close to physical exhaustion, as indicated subjectively and by abnormally high heart rates, by the end of their allotted time outside the spacecraft. Indeed, as a consequence of this heat overload, the objectives of the EVA were not completed during two missions.

For subsequent EVA missions, greater emphasis was placed on ground-based water immersion training to simulate work in microgravity, and design improvements were introduced to improve suit mobility, since much physical effort had been expended in working against the fixed resting position of the Gemini suit. Greater attention was also paid to the provision of restraint devices (hand holds and rails) in order to reduce the effort required solely to maintain position in space. The suit improvements introduced, for both free space (microgravity) EVA and lunar surface EVA, meant that metabolic heat production and elimination were no longer a problem during the Apollo and Skylab missions. It also became possible to calculate the metabolic rate during various phases of EVA using telemetered data from the Portable Life Support System (PLSS). The average metabolic rate for all activities on the lunar surface was $130\,\text{kcal.m}^{-2}.\text{h}^{-1}$, although driving the lunar rover was considerably less expensive with an average cost of $68\,\text{kcal.m}^{-2}.\text{h}^{-1}$. During the Skylab missions, EVA in free space also produced an average metabolic rate of about $130\,\text{kcal.m}^{-2}.\text{h}^{-1}$. This degree of constancy probably reflects the ability of the astronauts to pace their activities reasonably and considerably simplifies the design requirements of a PLSS.

The crews of the Vostok and Mercury spacecraft were obliged to remain within their capsules and in spacesuits throughout flight, and thermal comfort was achieved by the direct ventilating action of the ECS cooling sub-systems which supplied both the cabin and the suit via radiators (Vostok) or sublimators (Mercury). During the early EVA missions of Voskhod 2 and the Gemini programme, a similar open-circuit ventilation system was used to deliver pre-cooled oxygen to the suit, via the umbilical, for both breathing and temperature-control purposes. It had been deduced that a realistically achievable flow of about 300 litres per minute (L.min^{-1}) of suitably cooled gas would be able to remove up to 225–50 kilocalories of heat each hour (kcal.h^{-1}) by convection and evaporation, but the excessive build-up of heat seen during the Gemini 4, 9, and 11 EVAs showed this flow to be inadequate and additional methods of heat dissipation were introduced for subsequent flights.

Later Russian spacesuits have incorporated techniques for the removal of heat by evaporation of water contained in special panels within the suit or by conductive heat exchange with refrigerant circulating in tubes contained within a belt-mounted unit. The refrigerant is itself cooled in conventional on-board heat exchangers. In addition, the load placed on these cooling systems is minimized by careful choice of spacesuit layering materials. For the Apollo lunar EVA spacesuit, and indeed for most subsequent American suits, a liquid cooling garment has been incorporated, although some supplementary evaporative cooling was retained by ventilating at a flow of $170\,\text{L.min}^{-1}$.

A liquid cooling garment (LCG) provides cooling by conduction, as pre-chilled water circulates in polyvinyl chloride tubes mounted in a closely fitting undergarment. In the Apollo system, the tubes were stitched to the inner surface of an open mesh nylon layer, while a comfort layer of lightweight nylon chiffon separated the tubes from the skin. Cold water was delivered via a waist-mounted manifold to the garment either from the LM support system or, during EVA, from a cooling sublimator carried in the PLSS backpack. Warm water returning from the LCG entered the sublimator to be cooled before re-circulated. The system, which was first developed at the Royal Aircraft Establishment in Great Britain, was able to cope with large thermal loads for several hours. For example, very high metabolic rates ($278 \, kcal.m^{-2}.h^{-1}$) could be sustained for several hours without thermal stress while sweating at metabolic rates of over $200 \, kcal.m^{-2}.h^{-1}$ could actually be suppressed, and the average measured metabolic rate for all lunar activities ($130 \, kcal.m^{-2}.h^{-1}$) could be tolerated for eight hours!

The astronauts were able to control the suit temperature at any of three levels by means of a manual selection valve. For the early lunar surface expeditions of Apollo 11, 12, and 14, the temperatures were $21°C$, $15°C$, and $7°C$ but for the lunar roving EVAs planned for Apollo 15, 16, and 17 excessive cooling was prevented by increasing the minimum and intermediate temperatures to $27°C$ and $18°C$ respectively. The maximum LCG setting of $7°C$ was only used during periods of very high workload and no thermal problems were encountered at any stage of a lunar EVA.

Cooling for free space EVA from the Command Module during the later Apollo missions reverted to a gas ventilation system at a flow insufficient to sustain prolonged heat production greater than $140 \, kcal.m^{-2}.h^{-1}$, but no over-heating occurred of the sort seen in the Gemini flights, presumably because EVA was short and more thorough ground training had been undertaken. For the very limited but operationally vital EVA undertaken during the Skylab missions, primary thermal control was once again by means of a LCG with a capacity to remove heat to the same degree as the Apollo system. The garment was, however, supplied with coolant from the spacecraft via an umbilical and not from a backpack: again, no thermal problems were encountered.

There is no doubt that the use of LCGs not only achieves its primary purpose of maintaining thermal comfort but, in so doing, also reduces fatigue, minimizes dehydration, and enhances operational capability. These aspects will assume even more importance as the need for EVA construction techniques develop with the Space Station programme. The Shuttle EVA spacesuit incorporates many advanced design features and is fully described in chapter 8 but its method of temperature control retains the proven LCG technology.

Control of humidity

Humidity is the amount of moisture present in an atmosphere and is usually expressed as the percentage content of water vapour. When an atmosphere contains the maximum possible amount of water vapour it is said to be fully saturated and to have a humidity of 100 per cent (water vapour behaves as a gaseous constituent of the air and, as such, exerts a partial pressure: when the air has a humidity of 100 per cent, the pressure exerted by water molecules is termed the saturated water vapour pressure (SWVP)). Lesser amounts are expressed as relative humidities: that is, relative to the maximum amount possible for a given situation. Thus, a relative humidity of 50 per cent indicates that half of the maximum possible amount of water vapour is present in the air. The degree of humidity possible depends on a number of environmental factors including ambient temperature, atmospheric pressure, and air movement. A full description of the state of a given mass of air must include mention of all these variables, since the expression of humidity alone is meaningless.

The intimate relationship between humidity and human thermo-regulation was briefly mentioned in the previous section. Although variations in humidity have little effect on human comfort when temperature is being adequately controlled and thermal comfort being maintained, it does profoundly influence thermal balance when abnormal conditions prevail or when the limits of normal comfort are being approached. In the presence of a high humidity, the body's physiological response to over-heating is severely compromised and may be overwhelmed. It is because of this danger that humidity within spacecraft requires careful control. But in addition, there are several direct consequences of altered humidity upon human physiology. In this instance it is the absolute level of humidity, as expressed by the partial pressure of water vapour, which is of more significance. A low absolute value will lead to desiccation (drying) of exposed mucous membranes of the respiratory tract with dryness of the nose and throat, while inactivation of the protective cilia (hair cells, the wave-like movement of which carries debris away from the lower airways) of the bronchial tree can produce increased susceptibility to infection. Chapping of the lips and drying of the eyes and skin adds to the discomfort. High absolute humidity will not only compromise thermo-regulation, but will also encourage the growth of bacteria and fungi in moist skin folds with consequent infection. In order to avoid all of these problems, a partial pressure of water vapour of about 10 mmHg (1.33 kPa) is regarded as the optimum value for human comfort in a closed environment.

For the manned spaceflight programmes, the relative humidity within the pressure cabins has been controlled at levels between 30 per cent and 70 per cent (USSR) and between 40 per cent and 70 per cent (US). Over the range of cabin temperature chosen, these levels provided partial pressures of water vapour (i.e. absolute humidities) within the range of 3–19 mmHg (0.4–2.5 kPa). The principal problem is that of excess moisture in the air as a result of human metabolism, specifically from perspiration and respiration. Since the control of temperature is so intimately linked with that of humidity, production of moisture in the form of perspiration will be minimized if body

Table 6.1 Leading particulars of spacecraft environmental control systems

		Pressure (mm Hg)	Composition	Prebreathe	Oxygen source	Carbon dioxide elimination	Temp	Heat dissipation	Relative humidity
• RUSSIAN SPACECRAFT	Launch	760	21%O_2, 79%N_2						
	Orbit	760	21%O_2, 79%N_2	None	Chemical	Chemical (to <3.8 mm Hg)	11–25°C	Liquid-Air Heat exch Radiator	45–65%
	EVA	176–300	100%O_2	100%O_2 for 40–60 min at 760 mm Hg					
• MERCURY	Launch	760	100%O_2						
	Orbit	259	100%O_2	100%O_2 for 3 hr at 760 mm Hg	Gas	LiOH[2]	10–32°C	Sublimator	
• GEMINI	Launch	760	100%O_2						
	Orbit	259	100%O_2	100%O_2 for 3 hr at 760 mm Hg	Liquid + Gas	LiOH		Radiator/ (Sublimator)	
	EVA	191	100%O_2	None					
• APOLLO	Launch	760	Originally 100%O_2 then 64%O_2, 36%N_2	100%O_2 for 3 hr at 760 mm Hg					
	Orbit	259	95%O_2, 5%N_2	None	Semi-liquid/ semi-gas	LiOH (to <3.8 mm Hg)	24±2.8°C	Radiator + PTC[3] (+ evaporator)	40–70%
	EVA	191	100%O_2						
• SKYLAB	Launch	760	64%O_2, 36%N_2	100%O_2 for 3 hr at 760 mm Hg					
	Orbit	259	70%O_2, 30%N_2	None	Liquid	Molecular Sieve (to <5.0 mm Hg)	21–27°C	PTC (+ evaporator & radiator)	45–55%
	EVA	191	100%O_2						
• STS	Launch	760	21%O_2, 79%N_2						
	Orbit	760–527	21–26%O_2, 79–74%N_2	None or 100%O_2 at 760 mm Hg for 1 hr	Liquid	LiOH	16–32°C	Radiator	
	EVA	222	100%O_2	27%O_2 for 12 hr at 527 mm Hg, + 100% for 40 min					

Notes: 1 1mm Hg = 0.1333 kPa   2 LIOH = Lithium Hydroxide   3 PTC = Passive Thermal Control (rotation and/or paint)

temperature is maintained at a comfortable level by the cooling system. The cooling system will not be able to influence directly the other major insensible or 'hidden' source of body water: the water in the expired breath. Because of this, additional features are designed into the ECS to control the level of water vapour present. Such features must also be able to cope with sudden increases in humidity as a result of either increased muscular effort or increased ambient temperature.

For the ECS on board both Vostok and Voskhod spacecraft, relative humidity was controlled at 62 per cent to 69 per cent. An automatic sensor responded to variations outside this range of humidity by passing appropriately varying volumes of cabin air through a container filled with moisture absorbent. The system on board Soyuz craft differed only in that the air was dried by means of a heat exchanger rather than chemically. In all cases, manual control of the humidity-regulating system was possible.

In the Mercury spacecraft, water vapour was condensed during the passage of cabin air through the heat exchangers and the liquid allowed to collect in a water separator which contained a synthetic sponge. The sponge was periodically compressed to expel its water into a retaining tank. A similar system was designed for the Gemini programme: for the spacesuit sub-system, purified oxygen was cooled and dried in a combined heat exchanger/water separator, the condensate being carried to a collecting vessel by wicks, while a conventional heat exchanger and condensate collector provided the necessary water extraction for the cabin sub-system. The relative humidity within the Command Module of Apollo spacecraft was controlled within the range 40 per cent to 70 per cent using the same method. Water vapour was removed from the circulating atmosphere in the Lunar Module by means of two centrifugal water separators, but otherwise the system, and its design limits, were very similar. In the Skylab Orbital Workshop, overall relative humidity was controlled between 45 per cent and 55 per cent, although there were certainly local areas of the spacecraft with considerably higher humidities: on Skylab 4, one such moist area probably gave rise to the mildew found on a liquid cooling garment! Finally, the Shuttle ECS provides a thermal environment of 18°C to 27°C with an absolute humidity range of 6–14 mmHg (0.8–1.9 kPa).

With this discussion of humidity control, the description of the vital aspects of a spacecraft ECS is virtually complete. It will be readily apparent, however, that the overall control of atmospheric composition (oxygen, nitrogen, and carbon dioxide), atmospheric pressure, ambient temperature, and relative humidity within the closed environment of spacecraft or spacesuit is necessarily a single undertaking, although each aspect has been treated in isolation here for reasons of clarity. Table 6.1 summarizes the leading particulars of both Russian and American spacecraft environments.

Chapter seven

Nutrition, waste, and personal hygiene

Food and water

Without exception, the metabolic processes of life require energy for their accomplishment; but living organisms are thermodynamically unstable and as such will cease to function unless that energy is continuously provided from some external source. Furthermore, living organisms are also usually engaged upon a variety of other energy-requiring processes (that is, upon work), including that of physical activity, which is *the* most important variable determining total energy needs and, in the case of homeotherms, heat production. In man, energy for all of these functions is derived from the ingestion of adequate and appropriate foodstuffs. The three basic nutrients are proteins, fats, and carbohydrates (sugars); and all three must be consumed daily, along with small quantities of some minerals, trace elements, and vitamins, to maintain a satisfactory state of nutrition and health. In addition, water must be freely available.

Food

A seventy kilogram (eleven stone) man at rest has a basic daily energy need of 2,000 kcal (8,400 kJ) (1 calorie = 4.18 joules: both are units of energy). The additional energy needed by virtue of his daily activities may range from about 400 kcal (1,700 kJ) for light office work to over 2,500 kcal (10,500 kJ) for heavy industrial work such as mining. The caloric value of the food he ingests is therefore of great significance. For such a man undertaking light work, the daily intake of carbohydrates, fats, and proteins should be about 370, 65, and 70 grams respectively. Nearly 90 per cent of the energy needs are met by the carbohydrates (63 per cent) and fats (25 per cent), both of which are completely oxidized to carbon dioxide and water. Each gram of the former yields 4.1 kcal (17.2 kJ) and of the latter 9.3 kcal (38.9 kJ). Although fats have the greater yield, and are therefore the most compact energy source, it is the ease with which carbohydrates can be utilized which makes them the most useful and used source of energy. Proteins account for the remaining 12 per cent of the energy produced, each gram again yielding a caloric value of 4.1 kcal (17.2 kJ).

As well as providing a source of energy, the proteins and fats consumed must include certain essential components. Thus, at least 0.5 gram.kg^{-1} body weight of the protein should be of animal origin and contain the so-called essential amino acids (amino acids are the chemical building blocks from which human proteins are constructed, and essential amino acids are those which cannot be synthesized by the human body). Similarly, there are some essential fats, such as linoleic acid, which must form part of the overall fat intake. Common minerals, including sodium, potassium, calcium, iodine, and iron, must be present in sufficient quantity to meet the body's needs although a normally balanced diet usually contains more than is required; the fine control of body mineral content being governed by renal mechanisms. Minute quantities of trace elements such as aluminium, copper, magnesium, manganese, and zinc are also needed. Vitamins, which are organic substances essential to life by virtue of their catalytic action but which cannot be synthesized in the body, are the final components of an adequate diet.

Based initially upon intelligent estimations and later upon actual measurements of the levels of energy expenditure during various activities in space, and upon the well-established nutritional needs required by such levels on Earth, it was possible for the programme planners to devise suitable diets which met all the metabolic and constituent requirements. Additional constraints were imposed, however, because of the unique nature of the environment in which the food was to be eaten. Thus, in addition to the problems of actually eating in microgravity, any food sent into space had also to be stable with regard to the other aspects of cabin environment such as ambient pressure, vibration, temperature, and humidity. Furthermore, weight and bulk had to be as low as possible to minimize storage requirements. And, of course, the food had to be palatable with a pleasant appearance and texture, yet be free of pathogens (disease causing agents) and produce low residue and gas after digestion.

The energy requirements for both cosmonauts and astronauts have been progressively refined throughout the manned spaceflight programmes, largely as a result of in-flight experience and measurement. For Yuri Gagarin's short flight in Vostok 1, food was not essential, but he was provided with some puréed and liquid foodstuffs, which he successfully consumed, in order to establish man's ability to take sustenance under conditions of microgravity. At no time in either the Russian or American manned spaceflight programmes has the physiology of mastication (chewing) or deglutition (swallowing) been affected by the lack of gravity. For the remaining, longer, Vostok and Voskhod missions, a daily caloric intake of 2,600 kcal was provided. The early Soyuz cosmonauts were provided with 2,800 kcal.d^{-1}, but by the time Salyut 1 was launched this had been increased further to 2,950 kcal.d^{-1}; in Salyut 4 the figure was 3,000 kcal.d^{-1} and in Salyut 6 it was 3,150 kcal.d^{-1}.

A similar pattern has evolved during the American programme. Once again, the short duration of the early sub-orbital Mercury flights obviated the need for food, and neither Shepard nor Grissom ate at all during flight, but the nutritional needs for later Mercury missions, and for the Gemini and Apollo programmes, were based on a daily caloric provision of 2,500 kcal. The allowance was increased to 2,800–3,000 kcal.d^{-1} for the Apollo lunar landing flights. For Skylab too, with its intensive programme of physical exercise, an increase in caloric need was predicted and up to 3,000 kcal.d^{-1} were provided. This figure is also the energy provided by a typical daily menu on board Shuttle, although clearly no crew member is obliged to eat this much. For female adults in space, the daily caloric requirement is somewhat lower at 2,700 kcal and 2,000 kcal for Russian and American women respectively.

From the earlier discussion of energy requirements, it may be deduced that estimations of the level of activity while in a small space craft had the astronauts and cosmonauts expending little more energy than office workers. It has already been noted, however, that the energy cost of free-space EVA was initially underestimated and this, together with the larger habitable volumes of Salyut, Skylab, and Shuttle, the longer duration of space flights, and the increasing use of extensive in-flight exercise regimes, has led to the gradual revision upwards of energy needs in space.

For the early space voyagers, the manner in which this dietary requirement was qualitatively achieved left much to be desired, mainly because the constraints of weight, bulk, and ease of use were significant. Thus, the Vostok, Voskhod, and Mercury crews enjoyed their food as rather unappetizing purées and juices from squeezable aluminium tubes, or as vacuum-packed or freeze-dried bars. The food for Vostoks 1 and 2 was provided as 160-gram tubes of sterilized purée (such as meat and vegetables, meat and groats, processed cheese and prunes), and as similar tubes of various fruit juices and beverages (apple, plum, and currant juice, and coffee). Solid foods including bread, sweets, and sausage, all vacuum-packed in a synthetic edible film, supplemented the tube diet. For Titov's Vostok 2 mission, as for all succeeding Russian spaceflights a regimen of four meals a day was imposed based on Moscow time: breakfast, lunch, dinner, and supper. For the remaining four Vostok flights and the Voskhod missions, fresh fruit (apples and oranges) and bite-size portions of an extended range of foods was provided, including hamburger, roast beef, chicken fillet, fish, and caviar sandwiches. These rations were carefully prepared under aseptic conditions before being hermetically sealed and placed in one-meal packages. Tubes (156 g) containing puréed food (for example, meat and vegetables, and apples and peaches) were also provided for the Mercury astronauts. The purée was squeezed into the mouth through a short polystyrene straw (8.75 cm (3.5 in)), either directly or via an opening in the helmet if the facepiece was closed. Solid food, in the form of biscuit cubes, was assessed by

Carpenter during the second orbital flight: crumbs flew about the cabin and confirmed the need for packaging in edible film! Thereafter, a variety of compressed dry food cubes was provided, including fruit, bread, milk, and cereal. Rehydratable food was assessed for the first time during the twenty-four-hour final flight of Mercury 6 in May 1963, and this format was subsequently adopted for the principal method of provision in later programmes.

For missions of medium duration, such as the early Soyuz and Gemini and Apollo flights, an efficient food supply system was even more important. The cosmonauts on board Soyuz spacecraft, like their predecessors, were provided with diets composed primarily of natural, preserved, and hydrated food, since water-filled reconstitution devices were not carried. In addition to the tubes of purée and dried cubes of food, tin cans of a wide variety of meat products were supplied, including steak, chicken, veal, and ham. And, from Soyuz 9 onwards, a special heater was used to provide hot soups and drinks. A typical daily menu for a Soyuz cosmonaut is shown in Table 7.1.

The food supply systems of the Gemini and Apollo spacecraft were designed to support two or three men for at least fourteen days. About half of the food provided was in the more popular rehydratable form, water being delivered through a valve in the packaging from a central source. In the Gemini capsule and Apollo lunar module water at cabin temperature was used (21.1–27.6°C), but the Apollo command module offered a choice of hot (45.0–50.6°C) or cold (7.2–12.8°C) water. Solid food comprised the other half of the ration, and was supposed to be eaten while the main courses were being rehydrated (a process which took about ten minutes) or as snacks between the three principal planned meals of the day. An ever increasing range of foodstuffs was provided as the programmes progressed, so that eventually a

Table 7.1   Soyuz and Apollo: typical daily menus

Typical daily menu of a Soyuz cosmonaut	
Breakfast	Canned meatloaf, bread, chocolate sweets with nut praline, coffee with milk, prune juice.
Lunch	Canned beef tongue, bread, prunes with nuts.
Dinner	Caspian roach, bortsch, canned veal, bread, rich pastry, black currant juice.
Supper	Cream cheese with black currant purée, candied fruit, black currant juice.

Typical daily menu of an Apollo astronaut	
Breakfast	Apple sauce, sugar frosted flakes, bacon squares, cinnamon toast, cocoa, orange drink.
Dinner	Beef with vegetables, spaghetti with meat, cheese sandwich, apricot pudding, gingerbread.
Supper	Pea soup, tuna salad, cinammon toast, fruit-cake, pineapple-grapefruit drink.

four-day menu cycle was possible from the variety available. A typical daily menu for an Apollo astronaut is shown in Table 7.1

The food supply system was not without its disadvantages, however. The water used in the rehydration process for the Gemini and first two Apollo missions came from an electrical fuel cell and often produced an unpleasant taste as well as containing many bubbles of oxygen and hydrogen. Furthermore, the packets of food frequently ruptured during preparation and the contents which did remain in place required a great of time and effort to consume. For the remaining flights of the Apollo series, both the form and packaging of various items of food was improved. Sterilized, pre-packed meat dishes, with a high moisture content and capable of being eaten with a fork or spoon, replaced their dehydrated equivalents so that taste, texture, and ease of consumption were improved, and time was saved. Similarly, foods which still required rehydration were re-packaged more securely and in a manner which meant that they too could be eaten with a spoon.

Despite the apparent attraction of such menus, food consumption was frequently less than envisaged or desirable. Indeed, *all* the cosmonauts and astronauts on these pioneering missions lost weight, and, although this has been attributed to dehydration, both through diminished intake and increased loss by perspiration, the amount of food consumed must also be of relevance. For the long-duration missions of the Soyuz/Salyut series, the Russians quickly recognized the benefit of catering for individual preferences, and for many years now cosmonauts have been provided with personal menus. They have also been regularly supplied with fresh fruits and vegetables as part of the cargo carried to the Salyuts in Progress spacecraft. The American space diet has undergone considerable improvement as well, albeit to a slightly lesser extent. Over seventy different rehydratable food types were available for personal selection by astronauts on board Skylab. Thermostabilized food (that is, those which are stabilized by heating, such as pineapple, peaches, jams, and sauces), frozen foods (lobster, filet mignon, cakes, and ice cream), bite-sized dry and moist snacks (roasted peanuts, wafers, biscuits, and sweets) and several types of beverage (lemonade, coffee, tea, cocoa, and juices) supplemented the main food store. A communal table was designed for Skylab which incorporated water guns for rehydrating food, while individual meals were prepared on trays containing heating elements. And an even closer approximation of Earth-like gastronomy was achieved by providing an oven to heat food, and utensils with which to eat it through flaps in special, membrane covered dishes. The eating utensils were held in place by magnetizing the trays.

The Shuttle also has a fully equipped galley mounted on the floor of the mid-deck. The galley provides an area for the centralized preparation of meals and, to this end, incorporates hot and cold water supplies, a forced air convection oven (capable of heating food to 82°C) and a pantry, but no refrigerator or freezer. The pantry is stocked with a wide variety of foodstuffs both to supplement the main menu and to act as a contingency supply in the event of an unplanned extension to a mission. Both the pantry and the main supplies are constituted according to individual preference as discussed before flight (Plate 6). Pre-launch preparation of foodstuffs is similar to that for Skylab, and includes thermostabilized, irradiated, freeze-dried, and dehydrated types as well as natural form (biscuits and nuts) and fresh fruit and vegetables. Packaging, however, is continually refined in the light of experience in order to facilitate in-flight preparation and consumption. Table 7.2 shows some typical daily menus for a Shuttle crew.

Table 7.2 Typical daily menus for Shuttle astronauts

	Day 2	Day 3
Breakfast	Apple sauce (T) Beef jerky (NF) Granola (R) Breakfast roll (NF,I) Chocolate, instant breakfast (B) Orange-grapefruit drink (B)	Dried peaches (IM) Sausage (R) Scrambled eggs (R) Cornflakes (R) Cocoa (B) Orange-pineapple drink (B)
Lunch	Corned beef (T,I) Asparagus (R) Bread (NF,I) Bananas (FD) Peanuts (NF) Lemonade (B)	Ham (T,I) Cheese spread (T) Bread (NF,I) Green beans and broccoli (R) Crushed pineapple (T) Shortbread cookies (NF) Cashew nuts (NF) Lemon tea (B)
Dinner	Beef with barbecue sauce (T) Cauliflower cheese (R) Green beans and mushrooms (R) Lemon pudding (R) Pecan cookies (NF) Cocoa (B)	Cream of mushroom soup (R) Smoked turkey (T,I) Mixed Italian vegetables (R) Vanilla pudding (T,R) Strawberries (R) Tropical punch (B)

Notes: 1. B = beverage, FD = freeze-dried, I = irradiated, IM = intermediate moist, NF = natural form, R = rehydratable, T = thermostabilized.
2. Liquefied condiments (salt, pepper, mustard, ketchup, mayonnaise) are supplied in serving-sized packs.

Each main meal is pre-assembled in labelled pouches according to a six-day master menu, although the various components can be replaced or supplemented by items from the pantry. Servings are then prepared, where necessary by rehydration and heat, before being placed in compartments on a tray for consumption. Food packages are held in place in friction fit recesses, while utensils are restrained by small magnets. Velcro strips applied to the underside of the trays allow attachment to a table top or even to the astronaut's legs for dining. In all, it takes about half-an-hour to an hour to prepare a meal for seven on board the Shuttle.

As might be expected, the constituents of each Russian and American daily menu were carefully balanced so that a whole day's supply provided the appropriate quantities and types of proteins, fats, and carbohydrates (in the approximate ratio by weight of 1:1:3.7 respectively) and of minerals (particularly calcium, iron, magnesium, phosphorus, potassium, and sodium). A certain amount of fibre was also incorporated, although faecal residue was kept as low as possible for obvious reasons. Vitamin requirements for the Mercury, Gemini, and Apollo missions were met to American satisfaction by the provision of a balanced natural diet; but the Russians have always provided twice daily multi-vitamin pills, and the crews of Skylab were supplied with vitamin tablets containing the minimum daily requirement.

The Russian nutritional programme has also been modified occasionally by the addition of certain drugs believed to assist in protection against the detrimental physiological responses to microgravity. These aspects will be discussed further in Part 3, as will the behavioural considerations of eating in space, since the morale-boosting psychological importance of preparing, serving, and eating meals has long been accepted. Indeed, more attention is now being given to the visual presentation of food and the social rituals surrounding its consumption.

Finally, although we can survive without any food for many days (up to forty) provided that we have an adequate supply of water, emergency food rations have been carried on board all spacecraft for use either in the event of an unexpectedly prolonged mission, or as a supply for survival purposes should the craft land in a remote area. In Vostok craft, for example, the on-board emergency supply had a caloric value of 1,450 kcal.d^{-1} and took the same form as the primary food store, while the land survival rations were conventional dried foodstuffs with maximum energy yield, such as milk, cream, cheese, bread, chocolate, and sugar.

Water

Life on Earth could not exist without water, for it not only surrounds all living cells but also accounts for over two-thirds of their contents; and this is as true for complex multi-cellular organisms like man as it is for simple uni-cellular beings which must invariably live in water to survive. In the human body, this external environment of water is provided by the Extracellular Fluid (ECF), the first of two major water-containing compartments (shown in Figure 7.1).

The ECF surrounds all the cells of the body and provides an interface between them and the outside world. Across this interface all the nutrients necessary to sustain metabolism must pass to the cells, and all the waste products produced by metabolism must be carried away. The ECF accounts for about 27 per cent of the lean body weight (that is, about 19 L (33.4 pints) in a 70 kg (154 lb) man is water) and is itself composed of three compartments:

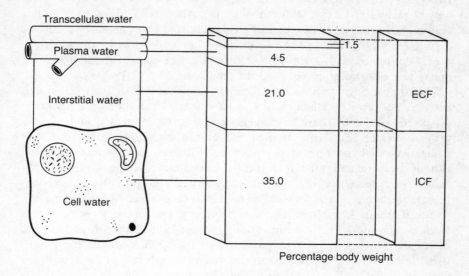

Figure 7.1 The fluid compartments of the body

the plasma volume of the circulating blood (4.5 per cent), the interstitial fluid surrounding the cells (21 per cent), and the so-called transcellular volume of secreted fluid contained within the gastro-intestinal tract, as digestive juices, and the central nervous system, as cerebro-spinal and intra-ocular fluid (1.5 per cent). The remaining body water, which accounts for about 35 per cent of the body weight (24.5 L (43.1 pints)), is located within the second major compartment and is termed the Intracellular Fluid (ICF). Most cells contain about 70 per cent water but fat cells are an exception and are relatively free of water. Thus, fat and water content are inversely related, and this accounts for the difference seen between water content in healthy young males and females: 62 per cent and 53 per cent respectively. In both sexes, water content declines with age.

When undertaking light work in temperate climates, about 2.5 L (4.4 pints) of water are normally lost each day from the body. Some is lost from the respiratory tract as water vapour in the expired air, and from the surface of the skin as sweat. This is termed the insensible loss and amounts to about 0.9 L (1.6 pints). A further 0.1 L (0.2 pints) is lost in the faeces, but the largest and most controllable daily loss is via the kidneys as urine: about 1.5 L (2.6 pints). Clearly, these figures will be modified dramatically by physical activity and climate (working in a desert, for example, may lead to a daily loss of over

10 L (17.6 pints)) but, whatever the magnitude of the loss, water balance can only be maintained if there is an equal intake. About 0.3 L (0.5 pints) of water may be recovered daily as water of oxidation; that is, as an end-product of body metabolism. The rest, however, must be regained as food and drink. Water contained within the former usually accounts for about 0.9 L (1.6 pints), while the latter makes good the remaining 1.3 L (2.3 pints).

This simple description is complicated somewhat by the inextricable link between body water, which acts as a solvent, and the substances (solutes) dissolved within it. These substances include salts, minerals, and organic proteins, which, by virtue of their concentration, impart to the water its colligative or physical properties, such as its freezing and boiling points, its vapour pressure, and, most importantly, its osmotic pressure. Osmosis is the spontaneous tendency of a solvent (such as water) to pass through a semi-permeable membrane into a solution of greater concentration (such as saline), and to dilute it. A semi-permeable membrane is one which prevents solutes from passing down the concentration gradient in the other direction. The increasing amount of solvent passing into the solution raises the latter's osmotic pressure until flow ceases: the fluids on each side of the membrane are then in equilibrium and their osmotic pressures are the same.

This is the situation which must obtain in living organisms in order to prevent cells either swelling up or shrivelling as a consequence of osmotic gradients. Thus the ECF and ICF are normally in osmotic equilibrium. Special receptors (osmoreceptors) in the hypothalamus of the brain are able to monitor changes in the osmotic pressure of the fluids surrounding and within them, and to respond to such changes by invoking hormonal alterations to which the kidneys respond. Under normal circumstances, this homeostatic mechanism is able to maintain water and salt balance with extreme precision. Furthermore, if a water load is taken, the dilute blood reaching the hypothalamus triggers a hormonal effect upon the kidneys which rapidly respond by excreting increased quantities of dilute urine. Excess body water is therefore seldom a problem in healthy individuals.

Similarly, in the presence of a water deficit, the kidneys are stimulated to retain water and to concentrate the urine. In addition, fluid intake is increased as a consequence of stimulation of thirst receptors within the hypothalamus. Correct balance is quickly re-established, provided that either the deficit is not great or water is readily available. If, however, water is not available, the deficit will quickly become severe since, even when the kidneys are retaining salt and water to their maximum ability, the insensible loss continues unabated and is usually increased. A clinical state of dehydration will then develop. Thirst may first be obvious when just 1 per cent of the body weight has been lost, but will become intense when the loss is 2 per cent. Appetite is lost as well and there may be feelings of discomfort and oppression. By the time the loss is 4 per cent, apathy, nausea, emotional instability, impatience, a flushed skin, and economy of movement may all be evident. Paraesthesiae, headache, and heat exhaustion, even in fit men, are apparent when 6 per cent of the body weight has been lost. And body temperature, pulse rate, and respiratory frequency will all be increased by this stage. Spastic muscles, a swollen tongue, delirium, and a general incapacity are seen at a weight loss of 10 per cent; and with progressive shrivelling, cracking, and numbness of the skin, failure of motor function, and failure of renal and cardiovascular function, death becomes almost inevitable when the weight loss exceeds 15 per cent to 20 per cent.

Dehydration may prove to be fatal within hours if circumstances are unfavourable, as in a desert, but people have survived fourteen days without water under ideal conditions! It is quite clear that an adequate fluid intake is an absolute requirement of all human endeavour, and drinking water has been freely available on board all manned missions into space.

For the relatively undemanding missions within spacecraft, the Russians have recommended a total daily fluid intake, as water and in food, of 2.2–2.5 L (3.9–4.4 pints); but these figures were doubled for those missions requiring EVA. Water for all Vostok, Voskhod, and Soyuz flights has been carried as a potable (drinking) supply prepared on Earth. Thus, because the taste of water deteriorates at room temperatures within eighteen to thirty-eight hours, even when stored in glass containers, such supplies have been treated with ionic silver which acts not only as a preservative but also as a sterilizing agent. The water was carried in sealed 5.0 L (8.8 pints) polyethylene film containers and enclosed in rigid metal cylinders. A short pipe connected a valve in the container to a mouthpiece which, when suction was applied by the user, allowed simple delivery of water. For Salyut 6 and subsequent craft, a 180 L (317 pints) water tank was installed. This replaced several of the small containers and was capable of replenishment from Progress cargo vessels. As with eating, no physiological problems have been encountered with the act of drinking in space.

For the short-duration missions of Project Mercury, the only water supply was that loaded before launch. Flexible pouches, with a capacity of about 2.7 L (4.75 pints), were used and drinking was accomplished by squeezing the pouch and forcing water along a flexible straw into the mouth. The power source for Mercury spacecraft was from conventional silver-zinc batteries but with the introduction of electrical fuel cells for subsequent craft, the on-board generation of water became a possibility. In such cells, hydrogen and oxygen are stored in separate compartments but allowed to contact each other through a catalyst. A chemical reaction then takes place which liberates energy, as electricity and heat, and produces water as a (convenient) by-product. The electricity is used as a power source, the heat is dissipated, and the water is used in other spacecraft systems. For drinking purposes, such water requires treatment with filtering devices and, since the fuel cells on Gemini craft did not have these filters, the Gemini astronauts, too, used stored water as their drinking supply. The daily recommended consumption standard of 3.0 L (5.3 pints) per man was provided from a single container, with a capacity of about 7.3 L (12.8 pints), stowed between the two ejection seats to supply both crew members. The main store was held in the adaptor section of the spacecraft and was used to replenish the cabin container until the adaptor was jettisoned for re-entry.

For the Apollo missions, both derived water and pre-loaded stores were provided. Three fuel cells, located in the Service Module, each provided 0.23–0.54 L (0.4–0.95 pints) hourly under normal conditions, and even more at times of high power generation. The water was cooled, passed through a gas-liquid separator which removed excess dissolved hydrogen, and then delivered to a control panel in the Command Module. From this panel, water was distributed first to the potable water tank and then, when this was full, to the waste water tank. The former had a capacity of 16 L (28.2 pints), the latter 25 L (44 pints) and any water in excess of the combined volume was dumped overboard. Both tanks were made of aluminium alloy, and a polyisoprene bladder within each was inflated with oxygen to maintain water pressure in the system at 1,295 kPa (25 pounds per square inch (psi)). Potable water passed from its tank to either a heater or a chiller, and thence to the food preparation unit for use in rehydration. Chilled water could also pass to a drinking gun designed to deliver water in aliquots of 14 ml (0.5 ounce). The potable water system also acted as a secondary means of temperature control via the spacecraft evaporators, the primary system being sublimation through radiators. A final link with the ECS was the delivery of water condensate to the waste tank from the humidity circuits of both the cabin and the spacesuits. Additional stores of water were carried to provide for increased crew requirements and as a contingency should the fuel cells fail.

Since the Lunar Module was powered by batteries, no supply of generated water was available and all potable water was loaded before launch. One storage container, with a capacity of 151 L (266 pints), was carried in the descent stage and supplied water for the landing and lunar surface activities (for the long lunar missions of Apollo 15, 16, and 17, a second such container was carried); while two 19L (33.4 pints) tanks were installed in the ascent stage for use during the recovery phases. For the lunar surface EVAs of Apollo 12 and subsequent missions, a 1.08L (1.9 pints) drinking pouch was installed in the spacesuit and this could be replenished from the main LM store. Finally, the water stores had also to provide the primary cooling for the LM through a sublimation process. Because electrical fuel cells could not supply sufficient power for the needs of Skylab, large solar arrays backed up with batteries were installed. This meant that no on-board generation of water was possible, and so over 2,900 L (2.9 tonnes) of water were pre-loaded and stored in ten cylindrical tanks within the OWS. A single 12 L (21.1 pints) portable tank was also carried. These stores provided water for all purposes, but potable water accounted for a possible 3.5 L.day^{-1} (6.2 pints.day^{-1}) for each crew member. Actual usage was only 75 per cent of this during the

Skylab 2 mission, but rose to 90 per cent for Skylab 3 and 4. Power on board the Shuttle is generated by fuel cells, and drinking water is provided from one of the two 75L (132 pints) tanks to which water generated by the cells is delivered. The other tank provides water for the thermal control system, and any excess is vented overboard.

As was mentioned earlier, it has been a consistent finding that virtually all cosmonauts and astronauts have lost weight during flight. Since this weight loss is made good within days of return to Earth, the phenomenon has been attributed to dehydration, rather than to a failure of caloric intake, despite the readily available supply of fluids on board spacecraft. There is certainly a diminution in feelings of thirst, and many crew members have had to be coerced into drinking an adequate amount. But the principal cause of fluid loss is by increased excretion of urine (diuresis) from the kidneys. The underlying mechanism of the diuresis is not fully understood but it may well be a direct consequence of exposure to microgravity and its effect on the body's homestatic mechanisms. This subject will be pursued further in chapter 11.

Management of waste products and personal hygiene

While life exists, the human body continues to produce large quantities of waste products as the end result of its metabolic processes. The principal waste products are carbon dioxide and water produced by the oxidation of foodstuffs: the breakdown is complete in the case of carbohydrates and fats, but proteins are incompletely oxidized and produce additional waste products in the form of nitrogen-containing compounds such as urea. Virtually all of the carbon dioxide, and it may be over 1 kg (2.2. lb) daily, is eliminated in the expired air, along with the obligatory insensible loss of a considerable quantity of water as vapour. The remaining daily water loss, and that of the nitrogenous wastes, is as sweat, and in urine and faeces. The faeces usually contain over 75 per cent of their weight as water, although total bulk clearly varies. Carried within this faecal mass, which amounts to an average daily elimination of 0.1–0.2 kg (0.22–0.44 lb), is 25–50 grams (0.87–1.75 ounces) of solid organic and inorganic material derived primarily from dietary residue, colonic bacteria, and cellular debris. Other cellular losses are continually taking place from the skin and its appendages (nails and hair) and these too must be regarded as human waste products.

During everyday life on Earth, the carbon dioxide and insensible water from the lungs and skin enter the atmosphere directly and are immediately and harmlessly incorporated into the ecological matrix. Under normal circumstances, and usually after appropriate sanitary treatment, urine and faeces also ultimately re-enter the biological chain. But the luxury of such dispersal is denied to the occupants of closed environments of spacecraft and spacesuits. It is abundantly clear, therefore, that the disposal of human waste products is a vitally important aspect of life-support engineering in space. Both the carbon dioxide and the water vapour produced by crew members are managed by aspects of the environmental control system already discussed; in Chapter 3 for the former and in Chapter 6 for the latter. This chapter considers the disposal of other body wastes, and especially that of urine and faeces: a perpetually bothersome aspect of life in space.

Management of urine and faeces

Quite sophisticated waste-management systems (WMS) have been installed in Russian spacecraft from the Vostok programme onwards (Plate 7). And, although the systems have undergone progressive improvement and refinement, their underlying pneumatic principle remains the same: as both urine and faeces are passed, they are drawn off by an air stream created by a waste disposal ventilator, rather like a vacuum cleaner. The air stream ensures that the waste products are contained within the system and do not contaminate the cosmonaut, his clothing, or his spacesuit. Urine was passed into a receptacle funnel while faeces were delivered into rubber containers. The system was designed to collect both simultaneously, and could be used even while a spacesuit was being worn, since entry to Russian suits is from the rear.

American sanitary arrangements were, at least initially, somewhat more basic. For the very early short-duration Mercury spaceflights, no specific facilities were provided for the collection and disposal of urine; an omission which caused considerable embarrassment to the first American into space during his four-hour delay before launch! For subsequent missions of the Mercury and Gemini programmes, and for the early Apollo flights, a personal Urine Transfer System (UTS) was used by unsuited astronauts. This system consisted of a roll-on rubber cuff (colour-coded for multi-crew flights!) which acted as a conduit through which urine was passed into a receiver and valve assembly, and thence either directly into the overboard dumping system or into a flexible collection bag with a capacity of about 1.2L (2.1 pints). If the collection bag was used for intermediate storage in this way, urine could be dumped overboard later.

The UTS was relatively simple and effective, but was not particularly hygienic or popular since it required intimate contact on each occasion it was used. From the Apollo 12 mission onwards, the transfer system was relegated to use as a back-up method when a new device, the Urine Receptacle Assembly (URA) was introduced. In use, this assembly was an open-ended cylindrical container which could be hand held; it was no longer necessary to use a penile cuff, and, although spillages were frequent, skin contact could largely be avoided. Urine was held within the receptacle, even in microgravity, by the capillary action of a hydrophilic honeycomb screen. From the receptacle, urine passed via a quick release fitting to the urine dump line at a maximum rate of 40 ml.sec^{-1} (0.07 pints.sec^{-1}), and the whole system could cope with a total volume of 700 ml (1.2 pints) during concurrent urination and dumping.

Both the UTS and the URA used the same method of overboard dumping via a long flexible hose connected to a dump valve on the waste management panel. When this valve was opened to the space vacuum, urine could be vented at a maximum rate of about 450 ml.min^{-1} (0.8 pints.min^{-1}). Filters were placed in the line to prevent clogging with particulate matter, and heaters were fitted to the dump nozzle to prevent ice formation.

For suited activity, on launch, during EVA, during recovery, and in the event of emergencies, the intermediate storage mode of the UTS was invariably used, but re-named the Urine Collection and Transfer Assembly (UCTA). In this case, the collection bag (with a capacity of 950 ml, 1.7 pints) was worn around the waist, and drained after urination either while the astronaut was still suited or after the spacesuit had been removed. The same dumping system was employed. A similar in-suit device was used on board the lunar module, although urine was subsequently held in a 8.9 L (15.7 pints) waste tank and not dumped on the lunar surface.

For some of the later Apollo missions, slight modifications to the urine sub-system were made to accommodate scientific and logistic requirements. On Apollo 14, a so-called Return Enhancement Water Bag was carried to provide an additional store of water (as urine) in case of a main water system failure. Such bags were also carried on all subsequent missions to pool twenty-four hour collections of urine in order to avoid overboard dumping at operationally or experimentally critical stages of flight: a haze of water particles tended to envelop the spacecraft for some time after dumping and this would have jeopardized some navigational and optical procedures. The stored urine was dumped in the usual way when convenient, although a single twenty-four hour sample from each crew member of Apollo 16 was returned to Earth for investigation of the fluid and electrolyte disturbances believed to have occurred during earlier missions. Finally, for every day of the Apollo 17 mission, each crew member collected his twenty-four hour urine output from which he then extracted a 125 ml (0.22 pints) sample for return to Earth and subsequent electrolyte analysis.

Faecal collection and disposal provided a considerably greater challenge, and the method employed for the Mercury, Gemini, and Apollo flights was regarded with universal distaste. When defecation was necessary inside the spacecraft, faeces were eliminated into a plastic bag taped to the buttocks; a manipulation which required considerable dexterity and an inordinate amount of time — forty-five minutes was one astronaut's estimate of the time required. Having placed the bag over the anal area, a finger cot was used to move the faeces down into the container. The bag was then removed and the area cleaned with toilet tissues which were then deposited in the bag. As if this was not bad enough, the astronaut had then to add the contents of a sachet containing a liquid germicide before sealing the bag and kneading the mixture in order to stabilise the faecal matter before storage and return to Earth. Each faecal bag, duly labelled, was placed in a second, outer bag (which also doubled as an emesis (vomitus) container) before being rolled into the smallest possible volume and placed in a waste stowage compartment.

The unpleasantness of the whole procedure was one reason why pharmacological manipulation of colonic function was adopted both before the flight, in the form of laxatives, and during the flight, in the form of drugs to reduce intestinal motility. In addition, a low residue diet was consumed for several days before, and throughout, a mission. Even so, soiling of skin, clothes, and cabin surfaces was a constant problem and the difficulties surrounding this natural body function present a less than romantic view of space travel! Whenever a spacesuit was worn for EVA, the innermost layer around the pelvis was a Faecal Containment System (FCS). This splendid title disguised the fact that the 'system' was simply a pair of absorbent underpants which would contain about 1 L (1.76 pints) of faecal material if defecation became inevitable while outside the spacecraft. Perhaps not suprisingly, there is no report that this facility was ever utilized. During intravehicular suited phases, the containment system was not worn.

The WMS on Skylab was a vast improvement. The daily urine output from each crewman was collected in a pooling bag and its volume noted before a single 120 ml (0.21 pints) sample was drawn off. These small samples were then frozen and stored for later analysis back on Earth, while the remaining urine was dumped overboard and a new collection started with a fresh bag. Passage of urine into the container was facilitated by a positive air flow, generated by a centrifugal liquid/gas separator, which carried the fluid into the management system where urine and air were separated before the former entered the pooling bag. Collection of faeces was similarly automated. Each elimination was collected in a separate plastic bag attached to the underside of a form-fitting commode. An electric blower device was operated by the user to carry faeces into the bag, which was then sealed, labelled, weighed, and placed in a vacuum drying processor. Since the plastic bag was permeable to gas but not to liquid, its contents were gradually desiccated and after about eighteen hours the dried faecal residue could be stored for subsequent analysis on Earth.

The Shuttle, too, is equipped with a sophisticated Earth-like WMS. It consists of a small compartment on the mid-deck, and is able to process and store urine and faeces from male and female crew members. Urine is collected in a cup and tube urinal, while faeces are eliminated into the commode. The device is used by sitting on the commode, while restrained at the waist and feet, and operating the toilet gate valve. This valve starts an electric motor which rotates a set of vanes and tines (prongs), collectively called the slinger assembly, which in turn entrains both urine and faeces in an air stream. The solid waste is shredded by the tines on the rotating vanes and is deposited on the sides of the commode chamber in a thin layer, where, once the gate valve is closed, it is dried by exposure to the vacuum of space through a toilet vent valve. The WMS is also able to process waste wash water, emesis bags, and general spacecraft rubbish, and to control the transfer of all waste materials to storage tanks. Overboard dumping of solid or liquid material is not routinely undertaken, although the facility exists to do so if necessary, but air and vapours from the storage units are vented regularly.

Other aspects of personal hygiene

Other, less obvious aspects of everyday human biology also assume considerable importance in a closed environment. Up to 8 g (0.3 ounce) of dead epithelial (outer layer) cells may be shed from the skin each day, and the atmospheric control system must include filters to deal with such debris. Furthermore, the health of the skin will deteriorate if rudimentary hygienic procedures are not undertaken. Skin disorders have been seen relatively infrequently, however, and only then under predictable circumstances. Thus, folliculitis (inflammation of the hair follicles) has developed in areas such as buttocks and thighs, where pressure, friction and moisture combine to macerate the skin. A similar phenomenon affects the face and neck after several days of enclosure in a helmet, while dermatitis has occurred in the skin underneath electrodes placed for physiological monitoring. It is important to minimize the chances of developing skin disorders, and the provision of skin care facilities has been progressively improved as mission duration increased. Similarly, the need to deal with hair and nail growth became more important as first weeks and then months were spent aloft. The hair on the top of the head grows at a rate of about 0.35 mm (0.14 in) per day (faster in women), and slightly faster on the chin. Operational constraints and hygienic considerations in space clearly determine that, for prolonged missions, the appropriate hair and beard style is short and shaved respectively. Hair trimming and shaving facilities must therefore be provided. Similarly, fingernails, which grow at a daily rate of 0.1–0.2 mm (0.004–0.008 in), will require trimming before their length causes interference with performance.

More serious consequences follow any lack of oral hygiene, with an abrupt deterioration in the condition of teeth and gums. The breath very rapidly becomes foul (halitosis), a thin deposit coats and discolours the teeth, and the gums become inflamed and bleed (gingivitis). Basic mouth care procedures are thus essential; and the use of a toothbrush and toothpaste has been found to be the most satisfactory method for use in space.

Interestingly, no regular facilities for skin cleansing, shaving, or oral hygiene were provided for the cosmonauts on the Vostok and Voskhod flights. For the Soyuz/Salyut series, however, full consideration was given to these aspects. Mercury, Gemini, and Apollo astronauts were equipped with small cloths for skin care, some moistened with an antiseptic (Hyamine) and some dry. Facilities for either electric or wet shaving were provided, although the latter proved to be the most popular. Cut hairs could either be collected by a small suction device, or could be contained within the shaving cream used with the wet razor, and discarded with the face cloth.

As well as conventional towels and numerous boxes of wet, dry, bacteriocidal, and general utility wipes, the Skylab astronauts enjoyed the additional benefit of a self-contained hand washer and a whole-body shower (Plate 8) for personal hygiene. The hand washer received and contained a spray of water, while the soap was attached to magnets. The shower facility consisted of a collapsible cloth cylinder about 2.04 m (6.7 ft) tall and with a diameter of 0.76 (2.5 ft). Straps within the cylinder allowed the user to remain in place while washing. Water from the on-board supply system was fed to a small unit with a capacity of 2.72 L (4.8 pints) in which hot water could be mixed with cold before delivery, with a liquid soap, to the hand-held shower head. Water was removed from the system by suction. After washing, rinsing, and removing excess water with a sponge, a warm air stream provided final drying. Sufficient quantities of water were carried on board to enable each crew member to shower once a week but, because the device had to be assembled for use on each occasion and required careful cleaning and drying before stowing, it was not used as often as planned. In contrast, the shower on board the second-generation Salyut space stations was a permanent installation and was used with relish. For the Shuttle crews, a hand-washing enclosure has been provided, but there is no shower.

On Gemini missions, oral hygiene was accomplished with chewing gum and a toothbrush, while small tubes of edible toothpaste, toothbrushes, and dental floss were provided for Apollo, Skylab, and Shuttle flights.

The human sense of smell is remarkably acute, and the vapours of many organic substances may be detected at extremely low concentrations. Occasionally, this olfactory performance may be useful in detecting and isolating equipment malfunctions, but body odours may be expected to be potential sources of annoyance. Up to 0.5 L of flatus may be released from the intestinal tract each day and this, combined with odours from sweat, may clearly be most unpleasant. The sense of smell rapidly acclimatizes, however, and it appears that the occupants of spacecraft may not be concerned by a level of miasm which is overwhelming to an unacclimatized individual! Notwithstanding this finding, body deodorants were supplied for use in flight. Additionally, the charcoal filters incorporated in the atmospheric control systems to deal with potentially toxic cabin contaminants, also cope with much unwanted organic vapour.

Finally, of course, living in space inevitably generates considerable quantities of non-human 'clean' waste, such as empty food wrappers and cans, residual foodstuffs, used towels, and equipment bags. On all spacecraft, therefore, provision has been made to stow such items. In small spacecraft, storage bags secured within the cabin or in special waste compartments sufficed. But on the long-duration multi-crew missions in large spacecraft, such as Skylab, rubbish was discarded in bags through an airlock into a capacious holding tank (on Skylab, the tank was the unused liquid oxygen vessel from the Saturn stage IVB). A compacting device for crushing cans was also tested on Skylab but took up to an hour to work. On the Shuttle, all rubbish is placed in bags before being stowed, in a compartment below the mid-deck floor, and returned to Earth.

All of the above hygienic considerations are not only essential for the physical health of those in space but also play a vital role in the maintenance of their morale, dignity, and general psychological well-being. Although these aspects are discussed in greater detail in Chapter 12, it is important to realize that humans seem to require both themselves and their surroundings to be clean and tidy. Indeed, lack of water for washing is regarded as one of the four most irritating aspects of prolonged existence in a small closed environment; the others being a sense of isolation, inability to move freely, and boredom.

Chapter eight

Dressing for space and mobility

Clothing

On Earth, clothing is worn primarily to minimize any potential loss of heat from the body, so ensuring that body temperature is maintained at a constant level. This most important role was mentioned in the earlier discussion on thermal control in chapter 6, while secondary roles include protection from external extremes such as very high or very low temperatures, rain, snow, dust, and other forms of possible contamination. Sartorial aspects are important but nevertheless subservient to these basic needs. Once in space, however, with its constantly controlled neutral environment, garments designed simply to allow the accomplishment of a given set of tasks are all that is required. Thus, clothing for space presents a somewhat different challenge to that of terrestrial attire, although both should be simple to use, convenient, and unrestrictive as far as possible.

A layered concept has proved to be, as on Earth, the optimum way of achieving the desired results during work and relaxation. For routine periods of flight, therefore, cosmonauts and astronauts have usually been equipped with an inner layer of underwear and an outer layer general purpose flight suit. A further, thermally protective, outer layer has also been available to cosmonauts for emergency use. During certain other phases, however, and particularly during EVA, additional protective requirements have been met by the provision of highly specialized clothing such as liquid conditioning garments and, of course, spacesuits. This chapter presents some details of these various items of apparel.

Underwear

Since underwear is in direct contact with the skin, it must, most importantly, be non-irritating. The material from which it is made has therefore to be inert but at the same time allow the free passage of heat convected, radiated, or evaporated from the body. It must also be light and elastic, have the minimum of seams and be relatively wrinkle-free. These last two requirements are particularly relevant for garments worn underneath pressure suits. Although cotton and linen have excellent properties and are hygienic, it has been found that their durability is improved by incorporating some man-made materials into the weave. Thus, underwear composed of knitted fabric has been provided for both Russian (cotton-rayon) and American (cotton fabric) crews. One example of the latter was known as a Constant Wear Garment (CWG) and was an all-in-one short-sleeved, footless suit with a special mount for the radiation dosimeter and a belt for the physiological monitoring system. The Skylab crews were also provided with a second type of short-sleeved underwear, called the Union Suit, made of cotton and with integrated socks. Finally, as described in the previous chapter, a specialized pair of underpants, the Faecal Containment System (FCS) was worn by the Apollo astronauts during lunar surface activities.

Flight suit

During routine flight, a general purpose flight suit is worn over the layer of underwear. Once again, this garment must be light, soft, comfortable, and both convenient and easy to use. It must obviously integrate well with the layers on each side of it and must meet certain ancillary requirements such as the need for pockets, a degree of fire retardation, and no impedance to thermal regulation. Some of these requirements are mutually incompatible since, for example, the inclusion of a high percentage of synthetic material will enhance fire protection but diminish the ability of the suit to allow heat exchange. A combination of natural and artificial materials has again been found most appropriate. And even the colour and colour intensity of the chosen fabric have more than aesthetic value in space because of the thermal influence they may exert: light blue has been the choice of NASA.

For the long-duration missions of Skylab, the classic flight suit could be replaced by a long-sleeved waist-length jacked woven from Durette material, and to which matching trousers could be attached. Furthermore, the trousers could be converted to shorts by unzipping the legs at the knees! Matching gloves and boots were also available. The Skylab wardrobe was completed by a supply of conventional and short-sleeved shirts and T-shirts. Since laundry facilities were not available, all necessary clothing for the entire time of occupation had to be stored in, and launched with, the Orbital Workshop (OWS). In all, fifteen modules containing various items of clothing sufficient for one man for twenty-eight days were carried.

A cobalt-blue, two-piece, cotton flight suit is worn by Shuttle crew members over a short-sleeved shirt and underwear. Pockets with zips and Velcro patches on the coveralls allow small personal items such as notebooks, pens, scissors, sunglasses, and torches to be secured while in microgravity.

Thermally protective suit

A quick-don thermally protective suit provides a final layer of protection for cosmonauts should their cabin heating fail, or should it be necessary to abandon the spacecraft in a cold environment at sea or on land. Clearly, the general requirements of such a suit are the same as for the other layers of clothing considered, but additional cold protection is best achieved by layering the suit itself. An outer layer of hard-wearing, waterproof material is lined with a wind-proof layer which should be impermeable to air as far as possible yet permeable to body water vapour. It must also be durable, light, and flexible. A warming layer, of sufficient insulation, comes next, and the whole is lined by a smooth, light, and durable inner 'skin'. The degree of thermal insulation provided will obviously be a factor of all the layers, but, in general, the higher the insulating properties of the warming layer, the thicker will be the suit. A compromise has, therefore, to be adopted. The garment used in the Russian space programme has aimed to provide an insulation of 2.0–2.5 clo, which is equivalent to that required for use in the spring and autumn on Earth. By comparison, the thermal insulation required in light summer is 0.5 clo, and for cold winter is 3.5 clo, while in the extreme Arctic the value is 5.5–6.0 clo. (The clo is the unit of clothing insulation: one clo is the amount of insulation which will maintain a normal skin temperature when heat production is 50 $kcal.m^{-2}.h^{-1}$ in still air at a temperature of $21°C$. It is approximately equal to the amount of insulation provided by a business suit in a temperate climate.)

Gloves, footwear, and headgear

While inside the spacecraft, gloves, socks, and shoes (if worn) must meet the overall requirements of the other items of clothing. Thus, they must be comfortable, non-irritant, light, and durable, with adequate thermal insulation. Generally, the same material as is used for underwear is appropriate for gloves and socks, although the former must not interfere with manual dexterity. Skull-cap type headgear is very convenient for mounting physiological sensors and communication equipment.

Liquid cooling garment

The liquid cooling garment (LCG), for use during EVA, was fully described in Chapter 6. It was worn next to the skin, beneath all other layers except underpants.

Spacesuits

Previous chapters have described certain specialized aspects of spacesuit design and function: Chapter 3 dealt with the atmospheric control system,

Plate 1 Untethered extravehicular activity. A Shuttle astronaut operates entirely independently from his spacecraft using a gas-propelled Manned Manoeuvring Unit.

Plate 2 Lunarscape. Twelve men have stood upon the Moon's surface. Here, an astronaut surveys the bleakness of the Taurus-Littrow area during the Apollo 17 mission.

Plate 3 Women in space.
Cosmonaut Svetlana Savitskaya
was the first woman to journey
into space a second time, and
the first to undertake
extravehicular activity.

**Plate 4
Extravehicular
Mobility Unit.** The
modular spacesuit
designed for Shuttle
astronauts
comprises twenty
major components
and is designed for
use with the
Manned
Manoeuvring Unit.

Plate 5 Personal Rescue Sphere. A fabric sphere inflated with oxygen provides a safe haven for stranded Shuttle astronauts. It can be carried to safety by a conventionally suited colleague.

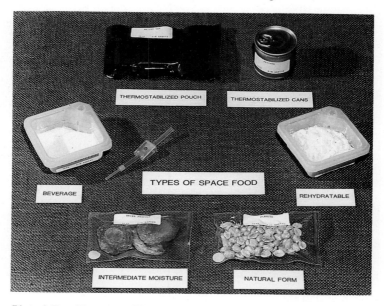

THERMOSTABILIZED POUCH

THERMOSTABILIZED CANS

BEVERAGE

TYPES OF SPACE FOOD

REHYDRATABLE

INTERMEDIATE MOISTURE

NATURAL FORM

Plate 6 Food in space. The choice of foods available to the space traveller is now wide and its presentation much-improved from the unappetizing puréed meals of early missions.

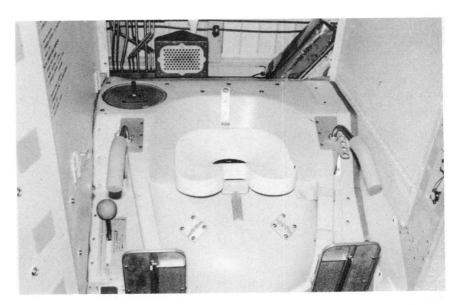

Plate 7 Mir toilet facility. The management of human waste in space has always been vexing. The private unisex lavatory on the Mir space station represents the best of modern technology.

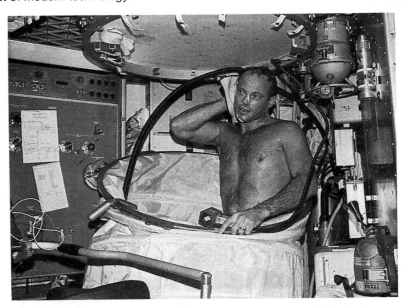

Plate 8 Skylab shower facility. The Skylab collapsible shower was provided with hot and cold water; once finished, the water was drawn off by a suction device.

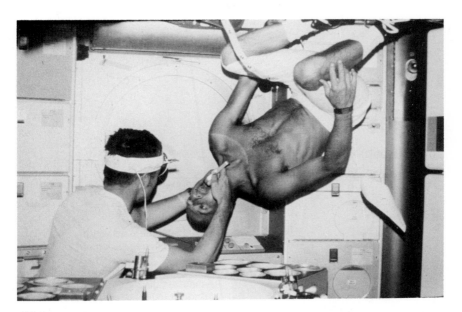

Plate 9 Dental care. Oral hygiene and dental health are vital during long space missions and all crew members are trained in basic dentistry: here being practiced on Skylab.

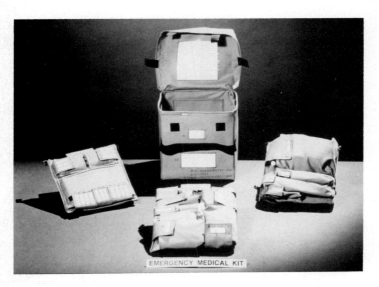

Plate 10 Shuttle emergency medical kit. Space travel carries the risk of illness or serious injury: kits are therefore always provided to allow the crew to deal with medical emergencies.

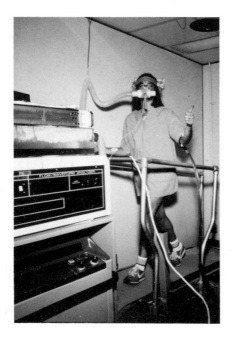

Plate 11 Medical selection. All candidates for space travel must satisfy many fitness standards, although these are no longer as stringent as those required for the pioneering cosmonauts and astronauts.

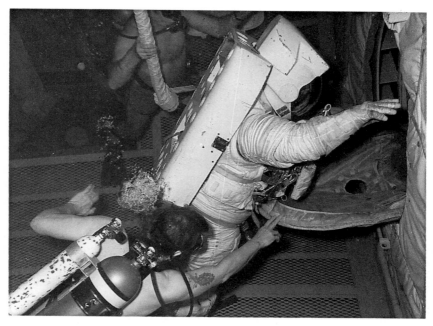

Plate 12 Training for microgravity. Simulation of extravehicular activity when underwater allows careful and detailed preparation for the work environment in space.

Plate 13 Microgravity. A unique environment which affects every aspect of human function – and can be fun.

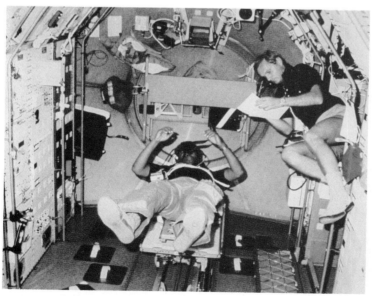

Plate 14 Physiology in space. Microgravity offers great opportunities to study the form and function of living organisms, including humans, with potentially enormous benefits for life on Earth.

Plate 15 Working in space. Techniques of construction in mircogravity must be refined to allow the building of even larger space stations which will act as stepping stones for deep space exploration.

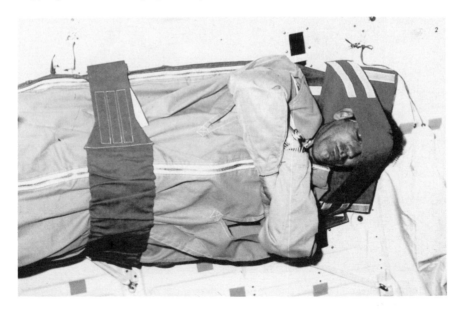

Plate 16 Sleeping in space. In microgravity, body orientation during sleep is irrelevant, but most people are more comfortable if some restraint is applied.

Chapter 6 with thermal control, and Chapter 7 with waste control. This section deals with the remaining structural considerations, and with certain supporting equipment.

A spacesuit is a pressurized micro-environment designed to protect its occupant against reduced barometric pressures at altitude. Jules Verne, in 1865, described just such a garment for use during his fictional journey *De la Terre à la Lune* (from the Earth to the Moon). As aircraft flew higher and higher so the non-fictional need for a pressure suit was identified and met, although it was not until 1934 that Wiley Post, the American pioneering aviator, made the first aircraft flight in a pressure suit. The technology of these full-pressure suits was refined in parallel with aircraft sophistication so that, in 1961, the Vostok cosmonauts and the Mercury astronauts wore suits almost identical to those of their colleagues flying high-altitude military jet aircraft.

In its most simple form, a spacesuit consists of just two layers: an inner rubber bladder designed to hold gas under pressure, and an outer restraining layer to prevent ballooning and so maintain shape. Clearly, the pressurizing gas must be supplied to the suit, as must the other vital environmental parameters such as breathable atmosphere, suitably cooled and humidified. And there must be arrangements for dealing with liquid and solid wastes. The combination of the spacesuit and its supporting supply systems comprise an individual life-support system; when components are entirely independant of the mother craft, that is, when a Portable Life Support System (PLSS) is used to supply the spacesuit, the whole ensemble is termed an Extravehicular Mobility Unit (EMU).

The use of acronyms is spacesuit technology does not stop here, however, and some further breakdown of parts may help to clarify the following discussion. The whole spacesuit is sometimes termed a Pressure Garment Assembly (PGA), while that part which encloses all but the head and hands is called the Torso-Limb Suit Assembly (TLSA). The head is enclosed in a Pressure Helmet Assembly (PHA), while the hands are covered by Pressure Gloves (PG). Other layers with specific functions may also have intriguing abbreviations, such as the Lunar Extravehicular Visor Assembly (LEVA), and the Integrated Thermal Micrometeoroid Garment (ITMG)!

Mercury spacesuits   The Mercury spacesuit was based on the United States Navy's full pressure suit as tested in the Man High balloon flights described in Chapter 2. It weighed approximately 10 kg (22 lb) and was a two-layer garment for emergency use should the primary cabin pressurization fail. Both layers were made of a nylon fabric: the inner was coated with neoprene rubber and the outer with aluminium to help resist the rise in cabin temperature during re-entry. The pressure helmet was a conventional enclosed dome of plastic with a visor of hard Plexiglass at the front. It was connected by a hermetically sealed heavy metal ring in the neck opening of the torso garment, and the whole helmet turned when the head was turned: it was thus a so-called rotational helmet.

Gemini spacesuits Because some of the Gemini missions entailed EVA, considerable changes were needed in the design and construction of the PGA to provide protection against the physical hazards of free-space activity and to enhance limb mobility. As a consequence, the basic Gemini garment, designated the G3C suit and based on an Air Force design, had an increased weight of 10.7 kg (23.6 lb) and had four layers: a white outer layer of high-temperature-resistant nylon (HT–1) capable of withstanding 500°C, a restraining layer of link-net material to prevent ballooning, a neoprene-coated nylon pressure-holding layer, and finally an inner layer of aluminium-coated nylon to provide protection against micrometeoroids and temperature extremes. For the first American EVA, by Astronaut White in Gemini 4, the protective qualities of the suit were enhanced by an outer cover which provided an additional seven layers of thermally insulating metallized nylon fabric (Mylar) sandwiched between four layers of micrometeoroid-proof HT–1 nylon. Special thermal gloves were also worn and the combined garment, weighing 18.2 kg (40.1 lb) was termed the G4C suit. The outer cover was dispensed with for subsequent Gemini EVAs but, in anticipation of the planned use of an Astronaut Manoeuvring Unit (AMU—a device to facilitate movement while in free space: see p. 121) during later flights, the backs of the EVA suit legs were coated with a temperature-resistant layer of stainless steel woven metal cloth (Chromel-R) to protect against hot gases from the unit.

The Gemini pressure helmet visor was initially made of Plexiglass but later of polycarbonate; and a second, removable, visor was added to shield the inner visor from impacts, and to filter the additional ultraviolet radiation encountered outside the spacecraft.

For periods of suited activity within the spacecraft, however, the Gemini EVA spacesuit was bulky and interfered with the ability to perform certain tasks. So for the fourteen-day mission of Gemini 7, the crew was equipped with a second, Intravehicular Activity (IVA), suit. This lightweight G5C suit weighed just 7.3 kg (16 lb) and had two layers: the usual inner neoprene-coated nylon pressure bladder and an outer temperature-resistant nylon layer. Link-net restraint was placed only over expansion areas. In place of the hard helmet, the suit had a fabric hood, with a polycarbonate faceplate, connected to the body garment by a long pressure-sealing zip. The zip could be partially undone to allow the hood to be folded back as a headrest, or could be completely released to allow the hood to be stowed. The boots were also removable and, for sleep, the whole garment could be doffed although one astronaut had to remain suited at all times.

Apollo spacesuits Project Apollo, with its requirement for EVA both in free space and on the lunar surface, saw yet more advances in suit technology, and the Apollo garment was a NASA development. Several versions were pursued, but the final suits, designated the A7L and the A7LB (with improved joint mobility), were used throughout the Apollo missions and reflected the lessons learnt from the Apollo 1 fire, particularly in the use of non-flammable materials. Again, a multi-layered construction was adopted but this time some of the layers were worn only during EVA.

The basic intravehicular suit PGA was worn during launch, rendezvous, docking, and recovery procedures and the TLSA, including the integral boots, was made of the same four functional layers as the Gemini suit. Apart from the neoprene-coated pressure layer, however, the materials were different: the outer layer was woven of Beta (glass fabric) cloth and was coated with Teflon, the restraint layer was made of fire-retardant Nomex, and the inner comfort layer was of a nylon material resistant to high temperatures. The transparent pressure helmet was made of polycarbonate plastic, with very high optical qualities, and had an internal foam pad bonded to the back of the shell to provide a head rest. It was attached to the torso by an aluminium neck ring. All inlet gas flow to the suit was eventually delivered to the helmet, where the foam pad acted as a manifold to direct gas flow to the front of the globe for respiratory and anti-misting purposes. This directional flow also helped to carry carbon dioxide away from the nose and mouth. In addition, there was a port through which food and drink could be delivered while in the spacecraft, even when the suit was pressurized. And it will be recalled that, for use during EVA, a drinking water bag was mounted in the neck rings of all suits from the Apollo 12 mission onwards.

The Apollo helmet was the first to be three-dimensional (cf. rotational) in that it was large enough to allow the head to move freely within it. The above-neck clothing was completed by a skull cap made of soft fabric. This closely fitting cap mounted the earphones, the microphone, and the associated electrical connections of the communications carrier assembly; the means by which the astronauts could talk to each other and to mission controllers. Glove assemblies were worn for both intravehicular and extravehicular configurations: nylon comfort gloves were worn next to the skin with a flexible pressure glove assembly on top, connected to the TLSA by a quick-release coupling. The Urine Collection and Transfer Assembly (UCTA) was also common to IVA and EVA suits.

Conversion of this intravehicular suit to the extravehicular configuration involved the use of an outer Integrated Thermal Micrometeoroid Garment (ITMG), an additional glove shell and the LCG. For lunar activity, the provision of a self-contained PLSS backpack was necessary, and the lunar surface EMU was completed by the FCS, the LEVA, and special lunar overshoes. The spacesuit of the Apollo EMU initially weighed about 29.5 kg (65 lb), but this increased to 31.8 kg (70.1 lb) when the PLSS duration was extended for the later missions.

The ITMG covered the TLSA and consisted of fourteen (!) layers of lightweight material designed to provide protection against the potential hazards of thermal extremes and micrometeoroids during EVA. The inner layer (next to the TLSA) was of rubber-coated nylon. On top of this were placed alternating layers of thermally protective aluminium-coated Mylar and spacing layers of non-woven Dacron: five of the former and four of the latter. Next came two layers of Kapton film/Beta marquisette laminate to provide further thermal insulation, and the final complete layer was of white Teflon-coated Beta cloth for abrasion and fire protection. Additional abrasion protection, in the form of Teflon cloth patches, was placed over the shoulders, elbows, waist, and knees; and metallized material covered the back where the PLSS was in contact. The ITMG mounted the pockets needed for the many small items carried by the space walker, while flaps provided for the routing of various connectors, such as the communication link, the physiological monitoring cables, and the urine transfer facility. For EVA use, a further, layered, glove shell with a gauntlet was worn as well, and secured to the inner pressure glove by Velcro fastenings. The shell layers provided the usual thermal protection, and also incorporated anti-abrasion protection in the form of a metal fabric (Chromel-R) woven into the palm and finger areas. Grip and tactility were improved by coating the palmar surface with silicone rubber and by making the thumb and fingertip shells of high strength rubber-coated nylon. The LCG, the PLSS, and the FCS have all been described previously, in Chapters 6, 3, and 7 respectively.

The Lunar Extravehicular Visor Assembly (LEVA) provided additional physical, thermal, and visual protection for the astronaut's head. It comprised a hard Lexan plastic shell which was worn over the pressure helmet and clamped to it, and on which were mounted three adjustable eye shades, one on each side and one central, and two visors. The outer, sun, visor filtered visible light and, by means of a thin coating of gold, rejected much infra-red and ultraviolet radiation; while the inner, protective, visor rejected infra-red rays and filtered ultraviolet.

Finally, the lunar surface overshoes provided abrasion and thermal protection for the TLSA boots while walking on the Moon. The soles were made of silicone rubber and had moulded ribs projecting from them to provide traction, rigidity, and additional thermal insulation. The rest of the boot was of a multi-layered construction of the same sort as the ITMG. The outer layer was of Chromel-R, however, and the inner of Teflon-coated Beta cloth. In addition, two layers of Nomex felt lined the soles as yet more thermal insulation. To summarise these complex ensembles, Table 8.1 lists the layers of clothing worn by Apollo astronauts during both intravehicular and extravehicular suited activities.

Skylab spacesuits The spacesuits used for EVA during the Skylab missions were virtually identical to the A7LB suits of Project Apollo, although

Table 8.1 Layers of the Apollo spacesuit

IVA	EVA
Skin	Skin
Constant Wear Garment	Faecal Containment System
	Liquid Cooling Garment
Pressure Garment Assembly	Pressure Garment Assembly
=	=
Torso-Limb Suit Assembly:	Torso-Limb Suit Assembly:
Comfort Layer (Nylon)	Comfort Layer (Nylon)
Pressure Bladder (Rubber)	Pressure Bladder (Rubber)
Restraint Layer (Nylon)	Restraint Layer (Nylon)
Cover Layer (Beta)	Cover Layer (Beta)
+	+
Pressure Helmet	Pressure Helmet
+	+
Pressure Gloves	Pressure Gloves
	Integrated Thermal Micro-meteoroid Garment
	=
	Inner Layer (Nylon)
	Thermal Layer (Mylar)
	Spacer Layer (Dacron)
	Thermal Layer (Mylar)
	Spacer Layer (Dacron)
	Thermal Layer (Mylar)
	Spacer Layer (Dacron)
	Thermal Layer (Mylar)
	Spacer Layer (Dacron)
	Thermal Layer (Mylar)
	Thermal Layer (Laminate)
	Thermal Layer (Laminate)
	Abrasion Layer (Teflon)
	Abrasion Patches (Teflon)
	Lunar Extravehicular Visor Assembly
	Glove Shell (layered)
	Lunar Overshoes (layered)
	+
	Portable Life Support System
	=
	Extravehicular Mobility Unit

obviously those components required specifically for the lunar surface were not worn. Thus, the role of the PLSS was replaced by supplies delivered to the astronaut by an umbilical from the spacecraft. A small chest-mounted unit, the Astronaut Life Support Assembly, similar to that worn by Gemini astronauts provided the necessary controls and warning devices.

Shuttle spacesuits The Shuttle EMU (Plate 4) is also similar in concept to that of the earlier programmes, but for reasons of cost the suits are no longer custom-made for each astronaut. Each EMU consists of a modular Spacesuit

Assembly (SSA) and a PLSS, and usually two such units are stowed in the airlock of the Orbiter ready for use by the nominated crew members. At least twenty major component parts make up the EMU, many of which are direct descendants of earlier versions including the PLSS itself, the LCG, the UCTA, the pressure helmet, and the extravehicular visor and communi-

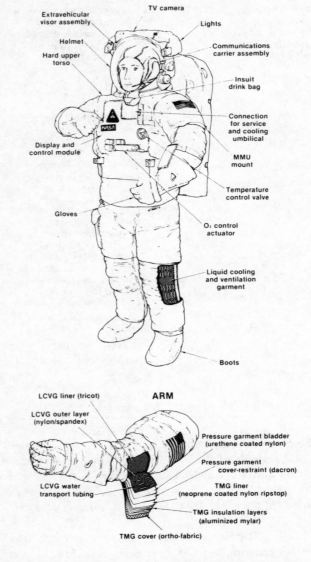

Figure 8.1 Schematic of the Shuttle Extravehicular Mobility Unit

cations carrier assemblies. A schematic of the Shuttle EMU is shown in Figure 8.1.

The main SSA, however, has been designed to allow interchangeability between various sized torso, arm, glove, leg, and boot portions. The fourteen separate components of the below-neck SSA are the hard upper torso and the lower torso assembly (five sizes of each), the upper arms (four sizes), the lower arms (thirty-six sizes!) and the gloves (fifteen sizes), the thighs (six sizes), the lower legs (eight sizes), and the boots (two sizes). Since all these items incorporate standardized joint connections, a suit can be assembled from any combination in accordance with the individual's measurements. Indeed, the full size range will accommodate the anthropometric spread of both male and female astronauts from the fifth percentile individual to the ninety-fifth percentile. (A percentile is one of ninety-nine values of a variable dividing a population into one hundred equal groups as regards the value of that variable.) The hard upper torso is made of fibreglass and has the PLSS permanently attached, but the rest of the suit is of the familiar layered construction, although the pressure bladder is of polyurethane and not neoprene. The helmet is made of blow-moulded polycarbonate, while the extravehicular visor assembly provides the usual impact, thermal, and visual protection. The entire EMU weighs approximately 110 kg (242 lb) and takes the wearer just five to ten minutes to put on (cf. the thirty minutes required to don an Apollo suit with assistance).

Mobility

Mobility and manual dexterity are obviously vital factors in the design of suits intended for use while working outside a spacecraft, and the large bulk necessarily involved in the layered garments described above does not help in this respect. Furthermore, as explained in Chapter 3, pressurization of the suit inevitably reduces limb and joint mobility even more. The constraints imposed by the requirements of suit pressurization and of protection against other external hazards have presented engineers with a most difficult set of problems. Nevertheless, the design of spacesuit joints has evolved and improved as the need for more extensive EVA capabilities has grown.

Artificial joints

Any artificial joint must clearly mimic the movement of the human joint over which it lies. Although some of the joints of the human skeleton can perform complex motions, most skeletal movements can be resolved into either hinge motion or rotational motion. Relatively full articulation of a spacesuit involves the mimicking of seven principal hinge-like joint movements: those of the wrist, elbow, shoulder, waist, hip, knee, and ankle. The two purely rotational movements of the wrist and hip, and the two combined hinge and rotational movements at the shoulder and hip must also be articulated. Technically, the success of artificial articulation depends upon the joint moving with minimum friction and minimum volume change.

For missions during which no EVA was required, there was clearly no need to provide spacesuits which were particularly mobile, and very little movement was possible in the early suits of Vostok and Mercury. Free-space EVA required considerable joint mobility, and the Gemini suits were equipped with joint systems which used link-net material to allow movement in a single plane. The joints were of non-constant volume, and were therefore physically demanding to use, since on flexing there is a tendency for this type of joint to return to its original extended position and work must constantly be done to overcome this.

The joints in the early Apollo suit (A7L) suffered from the same disadvantage and led to considerable restriction of movement and fatigue during EVA. Later, the principle of construction was changed to the use of fabric-reinforced moulded rubber convolutes (A7LB suit). A bellows-like stack of an appropriate number of these convolutes provided extension and flexion of a joint much in the manner of a concertina's action. Additional mechanical restraining devices were required around the complex joints (shoulders, waist, and hips) to help counteract the effects of loading from both suit pressure and natural movement.

For the Shuttle SSA, the joints of the hands and wrists, which require much dexterity, are of a pleated tucked fabric construction, while those of the other joints employ a constant volume, single axis, flat pattern joint with no 'spring back' tendency. Both the tucked fabric and the flat pattern joints are made of carefully sized pieces of material stitched together; the former being composed of simple lengths of fabric which tuck in, when flexed, like the folds of a fan, and the latter of gore (wedge) shaped sections which accommodate flexion without tucking. SSA mobility is completed by additional sealed ball-bearing joints at the waist, shoulder, upper arm, and wrist which provide considerable rotational capability.

Even greater mobility is possible in the most recent Russian spacesuit. This new lightweight suit was first tested on the Soyuz T2 mission and was used successfully during repair EVAs on Salyut 7. It weighs just 8 kg (17.6 lb) (cf. Gagarin's 20 kg (44 lb) suit) and is constructed of advanced plastics with highly mobile hinges at the knees, elbows, wrists, and fingers. Indeed, finger dexterity is such that a single match can be manipulated.

Aids to mobility

Several specialized aids to mobility have been provided for use by astronauts while undertaking free-space and lunar surface EVA. Once in the microgravity of free space, away from the security and assistance of hand rails and footholds on his solid spacecraft, a space walker has no means of either altering his own position in space or of moving from one point to another.

An artificial means of propulsion has therefore been required and a gas powered device, the Hand Held Manoeuvring Unit (HHMU), was designed for Project Gemini. This small T-shaped unit provided variable thrust through one forward-pointing and two rearward-pointing nozzles in response to hand pressure on a gun-like central control unit. For Gemini 4, the power source was from two small cylinders of oxygen mounted in the gun. Astronaut White very quickly used up the gas supply during his EVA and oxygen was replaced by a larger store of Freon, carried in the backpack, to power the HHMU for the intended space walk from Gemini 8. That mission ended prematurely and no EVA was performed. Perhaps surprisingly, Astronaut Cernan was not equipped with the HHMU for his subsequent space walk from Gemini 9. Despite the improved performance of his Extravehicular Life Support System (ELSS), the physical effort of the many tasks he had been asked to accomplish, including the use of the AMU, outstripped the ability of the ELSS to cope. The restraints provided on the spacecraft were hopelessly inadequate and Cernan became hot and very tired, his visor fogged, and the EVA was eventually abandoned. For the following two missions the HHMU was reinstated, although its power was now provided by nitrogen gas delivered from the spacecraft via the umbilical. The restraint devices were redesigned, but remained rather unsatisfactory, while the performance of the ELSS could not be improved because of time constraints. As with the earlier EVAs, the astronauts on these missions rapidly became fatigued as their ELSS failed to cope. So, for the final Gemini flight, yet more moulded footholds, waist tethers, and hand rails were provided, and pre-flight training in movement under water, to simulate microgravity, was intensified. Furthermore, the planned EVAs, by Astronaut Aldrin, were kept deliberately simple and relatively unambitious and indeed were, for the first time, the subject of extensive study in their own right: EVA had previously been merely the means to the accomplishment of another task. No HHMU was used on this mission.

In an attempt to provide a greater degree of mobility and to give the astronaut complete independence from his spacecraft, an Astronaut Manoeuvring Unit (AMU) was mounted in the Adaptor Module of Gemini 8 and 9. The AMU was a large rectangular aluminium backpack (81.3 cm high × 55.9 cm wide × 48.3 cm deep (32 × 22 × 19 in)) which housed life support supplies, communications facilities, and the hydrogen peroxide propellant for twelve small thrusters mounted on its corners. The astronaut was to sit in a form-fitting cradle and direct his own movements by means of controls mounted in two side arm supports. Once the AMU was released from the adaptor, only a 38.1 m (125 ft) restraint tether would connect him to the spacecraft. Unfortunately, the device was never used in free-space because

Gemini 8 returned to Earth before any EVA was performed, and the Gemini 9 EVA was discontinued when attempts by Cernan to don the unit merely resulted in exhausting him.

The lessons of Project Gemini as far as adequate spacecraft-mounted mechanical aids for free-space EVA are concerned were well learnt, and all subsequent spacecraft have been fully equipped with such devices. But for the lunar surface EVAs of Project Apollo, other mechanical aids were required. These included a full range of tools, a small wheelbarrow for transporting lunar material and, of course, the Lunar Roving Vehicle (LRV).

The folded LRV was transported to the Moon in a storage compartment of the Apollo 15 LM. Once on the lunar surface, the vehicle was unfolded by tugging on two lanyards and could be ready for use within fifteen minutes. The LRV was 3.1 m (10.1 ft) in length and weighed less than 200 kg (441 lb) on Earth. Each of its four wheels was driven by a separate electric motor, and open wire mesh tyres were used to minimize the bouncing effect of low lunar gravity. It provided transport for two astronauts, had a 9.65 km (6 mile) range of operation, and could cope with inclinations of up to 25° even when carrying its maximum load of 450 kg (992 lb). The radius of operation was in fact limited by the ability of the vehicle to return to the LM within the time provided by the emergency oxygen purge system; and this, therefore, imposed a limit of seventy-five minutes assuming an average speed of $8\,km.h^{-1}$ ($5\ miles.h^{-1}$) A dead-reckoning navigation system allowed accurate recovery to the LM and it was also equipped with very sophisticated radio and colour television facilities. The crews of the last three Apollo moon landing missions used the LRV with great success, the combined distance covered being just over 89 km (55.5 miles).

The most recent entrant to the catalogue of space mobility aids is the Shuttle's Manned Manoeuvring Unit (MMU). Prototypes of the definitive MMU, first used successfully during the flight of STS 10 in February 1984, were assessed by the Skylab astronauts eleven years earlier. Those trials were carried out inside the OWS, both with and without spacesuits, and over fourteen hours of experience were obtained. The MMU is a vital component of STS hardware since it allows the astronauts to work at considerable distances from the Orbiter on a wide range of activities, including satellite servicing, repair, and retrieval, space construction and experimentation, and, of course, rescue operations if necessary (Plate 1).

The MMU is really a very large backpack into which the astronaut, who is already wearing his PLSS, reverses. It consists of a central rectangular core made of aluminium, which houses two silver-zinc batteries, two nitrogen propellant tanks, and the duplicated controlling electronics. On each side of the core is a side tower and an adjustable horizontal arm: the former supports some of the directional thruster jets and astronaut displays, while the latter mounts hand controllers for thruster operation. The right hand governs rotational movement in pitch, roll, and yaw, and the left hand governs translational (straight line) movement in the x, y, and z axes. Twenty-four thruster units are arranged in two sets of twelve each fed from one gaseous nitrogen tank, although cross-feeding is possible should one system fail. Each jet can deliver 1.7 lb of thrust and the usual speed generated is about $0.6\,\text{m.sec}^{-1}$ ($2\,\text{ft.sec}^{-1}$). In addition, each tower has a fibre optic cable which runs forward to connect with the spacesuit display and allows the user to monitor the MMU's operation. Three location lights are also fitted and can be seen at distances of up to 1 km (0.62 miles). The whole unit is 125.4 cm high, 82.7 cm wide, and 120.9 cm deep (49.4×32.4×47.6in) when the arms are extended, and its fully loaded weight, including 11.8 kg (26 lb) of propellant, is 153 kg (337.4 lb). The total weight of a suited astronaut using the MMU may be as much as 346 kg (763 lb) on Earth.

When not in use, the units are stored in a special framework, called the Flight Support Station (FSS), behind the crew compartment in the forward end of the unpressurized cargo bay. From this position the batteries can be recharged and nitrogen replenished. Once a tethered astronaut has backed into the FSS and is held by its foot restraints, latches on the MMU connect with the PLSS. A lap strap is also fastened before the tether is released and the whole unit unlocked from its stowage position. Before releasing the foot restraints, however, the entire system is fully checked. Once satisfied that all is well, the astronaut can finally embark upon truly free EVA for up to six hours. The reverse procedure is adopted on return to the Orbiter.

Chapter nine

Health care

The preceding chapters have described the various physical and physiological hazards which confront all travellers into space; they have also attempted to explain the means by which many branches of science have combined to provide the necessary protection for such voyagers and so ensure their survival. All of this, however, presupposes that the individuals sent into space are, and remain, fully fit in body and mind. The physicians concerned with manned spaceflight are therefore obliged to provide the requisite medical care before, during, and after flight, so that health is maintained and, with it, the ability to perform at peak efficiency. The health care of cosmonauts and astronauts is thus accomplished largely by the application of preventive medicine, for if illness or injury occurs in space the success of a mission and indeed the very survival of the crew members may be seriously compromised. This chapter deals with the main aspects of medical care at all stages of a space mission, including the medical criteria determining crew selection. Apart from relatively minor differences in concept and detail, the Russian and American approach to this subject has been similar.

Pre-flight medical care

The most obvious aspect of pre-flight medical attention is the screening and selection process, but once this is completed an aggressive policy of increasing observation, intervention, and even semi-isolation accompanies the build-up to a launch in an attempt to forestall any medical deterioration.

Medical selection

The most obvious aspect of pre-flight medical attention is the screening and not only physically and mentally capable of performing the required tasks in space but are also able to withstand the stresses of spaceflight without being incapacitated. Furthermore, in the case of professional cosmonauts and astronauts (as opposed to scientist-astronauts and payload specialists), a clear corollary to this primary aim is that there should be no underlying physical ailment likely to preclude the successful completion of an entire career in the cadre if necessary.

The medical selection of the early Russian cosmonauts was a continuous process and involved three stages. During the first, candidates were screened for obvious contra-indications to the stresses of spaceflight, particularly acceleration, vibration, noise, microgravity, and isolation; and a great many were rejected because of pre-existing eye and ear, nose, or throat (ENT) problems. Internal disorders and neurological disturbances were among other reasons for failure at this stage. A highly detailed physical examination formed the second stage, during which an active search for latent (hidden) pathology was undertaken. Extensive tests of the cardiovascular and endocrine systems (because of their role in the response to stress), of the gastro-intestinal system (because of the propensity for latent disease of abdominal organs), of the neuro-vestibular system (because of its response to increased and decreased accelerations), and of fluid and electrolyte homeostasis (because of the importance of orthostatic stability) were performed. The spinal column was also subjected to detailed study, and so-called load tests of physical capacity and of the response to other stresses, particularly hypoxia and thermal insults, were carried out. Not surprisingly, all these procedures resulted in a number of further rejections primarily on the grounds of internal disorders, ENT problems, congenital anomalies, and degeneration of the spinal column. Assuming that the individual was satisfactory from other respects, success at this stage was followed by a period of professional cosmonaut training intended to increase ability to withstand the rigours of spaceflight and to improve physical and mental efficiency during the mission. The third and final phase of the Russian selection process thus involved possible rejection because of physical or, more probably, behavioural problems associated with the training programme. As experience was gained, the selection process was evolved so that the first, screening, phase became more demanding and rejections at later stages subsequently declined.

In the United States, a similar staged process was adopted for the selection of the 'Original Seven' Mercury astronauts in 1959. In fact, there were only supposed to be six, but apparently none of the seven could be omitted from this final choice. These seven had passed through an initial screening of medical and other records along with over 500 others from all branches of the American Armed Forces! Further scrutiny by NASA reduced this number to 110, and eventually sixty-nine candidates were invited to Washington, DC, for briefings, interviews, including a psychiatric assessment, and written tests. The first group of thirty-five yielded thirty-two suitable volunteers for further assessment and so the remaining pilots were not called forward. These thirty-two highly qualified professional men then underwent the second, active, phase of selection, part of which involved an incredibly detailed array of medical procedures. This assessment occupied seven and a half days at the

Table 9.1 Medical selection procedures for American astronauts

Full medical history and review, including flight experience

Full physical examination, including especially:
- central and peripheral nervous systems
- heart and lungs
- sinuses, eustachian tubes, and larynx
- eyes and ears
- teeth
- abdomen, including procto-sigmoidoscopy (internal examination of rectum and bowel) and full surgical evaluation

Special cardiac examinations, including:
- ECG examination at rest, during stress tests, hyperventilation, and breath-holding
- vectorcardiography
- phonocardiography
- tilt-table studies

Pulmonary function tests

Vestibular function tests

Special ophthalmic and aural tests

EEG examination

Laboratory studies, including:
- haematology
- serology
- liver function tests
- faecal analysis
- urinalysis

Psychological examination including intelligence tests, aptitude tests, personality profiles, and performance tests

Radiology (X-rays) of: skull, teeth, chest, spine, gall bladder, and entire upper gastro-intestinal tract (and colon if indicated)

Stress testing, including:
- exercise tolerance (Treadmill and Double Master's test)
- thermal insults (hot and cold)
- acceleration (centrifuge)
- decompression, including use of pressure suits
- isolation and confinement

Lovelace Foundation Clinic in New Mexico, followed by extensive stress (load) testing at the United States Air Force Aerospace Medical Laboratory in Ohio.

Very similar medical evaluation of subsequent groups of candidates was carried out over the next ten years at the United States School of Aerospace Medicine in Texas, with validation by NASA at the Manned Spaceflight Centre (MSC) in Houston (Plate 11). During this time, six further pilot/astro-naut/scientist applicant groups were processed, from which a total of sixty-six men was selected to join the astronaut corps. Table 9.1 lists some of the tests and procedures carried out on these men.

The extent of this examination was particularly remarkable in view of the fact that many of the candidates were already aircraft test pilots with the stringent medical attributes that that occupation requires. Because of this, and because space medicine was in its infancy, it was often very difficult to decide quite what would or should be disqualifying. A system of ranking candidates against each other was adopted rather than a straight pass or fail result; and such ranking even involved some peer group assessment.

The two groups of scientist-astronauts, recruited by NASA in 1965 and 1967, were subject to very much the same test and selection procedures as their pilot colleagues, but the severe stress testing was abbreviated for the second group in the light of experience gained. The scientist-astronaut corps was derived from the academic scientific community and, predictably, many candidates failed to meet the necessary medical standards. Equally marked was the comparison between the results of psychiatric and psychological testing of this group with those of the military-trained pilot-astronaut cadre. Both groups were evaluated using standard psychological procedures, and neither group demonstrated any significant psychiatric tendencies. The pilot-astronaut was typically very healthy, resilient, and intelligent; with a great interest in, and a high motivation for, flying and mastering increasingly complex vehicles. He was also well-organized, pragmatic, and aggressive in a non-hostile, ambitious way. Confident in his own abilities and not given to introspection, but with a subtle sense of humour, he nevertheless dealt with inter-personal relationships in a distant manner. These are essentially the recognized characteristics of professional test pilots, and are in no way representative of the general population of pilots, whether military or civil! The typical scientist-astronaut was also extremely fit and highly intelligent, but was driven by somewhat different ideals. Spaceflight, although intrinsically interesting, was to him a means to an end: the means by which his research could be advanced. Thus, there was not usually any driving ambition to travel into space: rather, his motivation stemmed from what knowledge was to be gained once there. The psychological tests revealed impatience with routine procedures (a trait incompatible with a test pilot's profession), a tendency to be more openly aggressive, and a resolve to complete a task come what may once embarked upon a set course of action, albeit wrong. The scientist did, however, score consistently higher than the pilot in tests of verbal, mathematical, and engineering achievement.

The conclusion of Projects Apollo and Skylab, and the subsequent decline in the American manned space programme, meant that there was no further requirement for new astronauts for some years. By 1976, however, NASA was mindful of the need to recruit for the forthcoming STS missions, and of the likely fundamental change in crew composition. A formal set of medical evaluation criteria was developed which reflected the individual requirements of different categories of crew members, and which allowed some tailoring of selection. In addition, it was decided that medical assessment would be conducted at the MSC (now called the Johnson Space Centre). The new criteria were published in 1977 and distinguished, largely on the basis of visual requirements, three classes of space personnel: Class 1 was the Pilot Astronauts, Class 2 the Mission Specialists (formerly scientist-astronauts), and Class 3 the Payload Specialists. Furthermore, much of the physiological stress testing for pilots and mission specialists was transferred to the one-year training and indoctrination phase, prior to formal acceptance as astronauts. Payload specialists, however, received just three to eight months' training.

The NASA call for recruits for the Shuttle programme in 1978 and 1980 yielded over 11,500 applicants from around the world. Of this total, 329 were called forward for further assessment using the new system and eventually fifty-four astronauts, including eight women, were selected for pilot (twenty-three) or mission specialist (thirty-one) status. Medical reasons accounted for 33 per cent of rejections, with visual problems alone responsible for 34 per cent of these. Psychiatric and cardiovascular problems accounted for a further 29 per cent. Interestingly, but not surprisingly, just 11 per cent of those rejected were from the candidates for pilot status. Separate arrangements have been made for the recruitment of various payload specialists from industry, for Spacelab, for other missions involving international co-operation, and for military flights. The last have also required the training of additional mission specialists.

Health Stabilization Programmes

Although it had always been recognized that astronauts nominated as prime crew members were at great risk of acquiring Earth-bound ailments during the immediate weeks and days before a flight, it was not until the operationally significant medical events leading up to the Apollo 8 (gastroenteritis), Apollo 9 (respiratory tract infections), and Apollo 13 (exposure to rubella) missions that the vital importance of pre-flight medical care was fully appreciated. For all subsequent Apollo and Skylab missions, the astronaut corps in general, and those of its members destined for immediate flight duties in particular, were subject to the Flight Crew Health Stabilization Programme (HSP). This ultimately most successful programme aimed to prevent any decline in an astronaut's health before, during, and after a mission. Of particular importance, however, were the three weeks immediately preceding flight during which communicable diseases may be acquired but remain covert until the victim is in space.

The first element in the HSP was the provision of a continuous and intensive clinical medicine service for all astronauts and their immediate families. This general practice concept provided day-to-day routine and emergency care, and fostered confidence and continuity. At a more active level, immunization against a large number of bacterial and viral illnesses was performed, although the most likely ailments, those of the upper respiratory and gastro-intestinal tracts, were and are not yet amenable to such prevention. In those astronauts and their children who were not already immune, routine immunization against diphtheria, mumps, polio, rubella, rubeola (measles), smallpox, and tetanus was administered. The astronauts also received influenza, typhoid, and yellow fever vaccines; and their children received protection against pertussis (whooping cough). Tests for syphilis and tuberculosis completed the barrage of immunological investigations for the intrepid astronaut!

No matter how effective all of these active preventive measures were, however, exposure to disease could still have negated their aims. Absolute isolation of the crew for the vital three weeks before a launch would have provided one solution, but this was obviously unacceptable from an operational standpoint, if not a social one. Other active measures were therefore adopted. For that crucial twenty-one days, only those people with whom it was essential to be in personal contact were allowed near the astronauts; and the process of eliminating possible carriers of disease began with a medical surveillance programme three months before the flight. Two months before launch, key personnel were questioned and physically examined and anyone considered to have, or to have been in contact with, a transmissable illness was excluded, as were uncontrolled casual contacts such as visitors and children (a particularly high risk group). Immediate adult family members were, of course, regarded as essential contacts although the children of astronauts were excluded. Usually, a total of about 100 to 150 people were permitted close to the flight crew. Living and working areas were rendered as safe as possible by the use of bacterial filters in air supply ducts and by positive pressure supply systems for air conditioning; while food and water were carefully procured, prepared, and tested under the direction of the medical team.

Interestingly, the Russians adopt a much abbreviated version of this programme, and are content to limit contacts for just one week before flight, to relieve personnel who become unwell or are overtly so, and to rely upon normal disinfection and hygienic procedures to prevent illness in the cosmonaut population. It is, however, standard practice for cosmonauts to bathe in alcohol on the night before launch and to sleep in beds sterilized by ultraviolet light; personal contact thereafter is kept to a minimum.

For the STS too, the HSP has been modified and three levels of care now exist. A Level 3 health care protocol would impose a virtual quarantine on the crew and has not been used so far. Level 2 most closely approximates to the original programme, but the period of minimal controlled contact is reduced, and this level was adopted for the STS-1, flight. Arrangements for subsequent missions were downgraded to Level 1, which is the least stringent. In Level 1, maintenance of health within the community relies to a large extent upon improved education and awareness, and both surveillance and examination are kept to a minimum.

Although the early astronauts were carefully examined at regular intervals in the period before their flights, it was not until the implementation of the

HSP for the crew of Apollo 14 that an extended pattern of pre-flight medical examinations was adopted. A thorough history was taken and a general examination performed twenty-seven days before launch (L–27) to allow time for any necessary preventive or therapeutic measures to be instituted. At L-15, an interim examination, including a dental inspection, was performed. And ten days later (L-5) a fully comprehensive examination, including cardiac, visual, aural, neurological, and radiological assessment, was carried out. Brief examinations took place daily thereafter, and final readings of vital signs, height, and weight were obtained at L-0 to provide a baseline for in-flight and post-flight comparison. For the Shuttle crews, the number and intensity of the pre-flight medical examinations has been radically reduced so as to minimize the disruptive effect on training. Four such examinations were carried out for the OTF pilots, and this number has been halved for routine STS missions. All crew members who have not been into space before do, however, undergo a vestibular assessment some six months before flight so that individuals susceptible to motion sickness can be established on effective treatment regimes. A similar protocol is adopted for experienced astronauts who have suffered from motion sickness on previous flights.

In-flight medical care

The health care of cosmonauts and astronauts while in space revolves around the ability to monitor their vital signs and to be able to act appropriately should a medical problem develop. To a great extent, these aims have been achieved objectively by ground-based physicians watching and acting upon information provided from space by biotelemetry. Of course, the crew members themselves are able to volunteer vital subjective information as well, and any deterioration detected has then been the subject of combined discussion and, if necessary, subsequent intervention by the men in space. The presence of a physician-astronaut on board is quite obviously advantageous, since it permits the expansion of medical tests to include such things as measurement of arterial blood pressure and pulmonary gas exchange. Diagnosis and treatment are also enhanced.

As described in Chapter 2, the physiological behaviour of early cosmonauts and astronauts was monitored as a matter of routine. Of particular importance was the ability to monitor indications from the cardiovascular system (as heart rate), the respiratory system (as respiratory frequency), and the neurological system (as level of consciousness). Heart rate was conveniently obtained from chest-mounted electrocardiogram (ECG) electrodes which also provided other information concerning cardiac function, while respiratory frequency was obtained from a device worn around the chest and which responded to chest movements by altering its impedance. Although some cosmonauts wore electroencephalogram (EEG) electrodes attached to the scalp to provide the necessary neurological data, the Americans relied upon simple voice communication, combined with television pictures, to provide an indication of mental health! Some other variables, such as oxygen usage and thermal loads, were monitored during lunar surface activities in order to assess the metabolic cost of work. And while very many other functions have also been studied, these have largely been concerned with in-flight medical experimentation rather than with primary health care. (A description of these experiments and their results form a large part of Chapter 11). The results of the prophylactic physiological scrutiny of man in space, however, revealed that the insult of space travel interfered very little with normal function, and the crews of later missions have gradually been able to dispense with such intense routine instrumentation.

Throughout the Mercury, Gemini, Apollo, and Skylab missions, a Bioinstrumentation System (BIS) was used to provide instantaneous data for physicians on Earth, except when Apollo was on the dark side of the Moon. The BIS was progressively improved and refined and all crewmen until Apollo 14 were instrumented with it. Thereafter, for the remaining flights of Project Apollo, just one crew member was monitored continuously, although the BIS was used for lunar surface activities. The Skylab astronauts also wore the BIS for EVA and during other important, operational, suited phases. Not surprisingly, the pilots of the Shuttle OTFs wore the BIS during critical stages of their missions, and ECG data were recorded from the mission and payload specialists during the launch and recovery phases of the early operational Shuttle flights, but continuous monitoring is now performed only during EVA. The BIS for Projects Mercury and Gemini incorporated two ECG systems, with impedance pneumography for respiratory frequency. One ECG system was deleted for the Apollo BIS which, apart from some improved electronics, was identical to its predecessors. The ECG was a two-lead device with one electrode on the sternum and the other in the axilla; while two more electrodes placed on the chest provided the respiratory recording. The electrode leads were routed to a so-called biobelt, within which were mounted the signal conditioners and other electronic units. From the biobelt, which was attached to either the CWG or the LCG and worn beneath the spacesuit, information passed via an umbilical to the spacecraft telemetry system and thence to Earth, sometimes from a distance of over 250,000 miles (402,250 km)! The data were presented in digital and analogue form to doctors at consoles in the mission control room. The BIS performed reliably, although displaced electrodes were a recurring problem, and the information passed back to Earth allowed mission physicians to adjust activities in space to the physiological responses seen. This was particularly important during EVA phases, the metabolic cost of which had not been predicted accurately prior to spaceflight (see Chapter 6). Biotelemetry also provided a potential source of clinical information which, fortunately, was not often tapped. Even the cardiac dysrhythmias noted during the Apollo 15 mission did not require treatment, although subsequent spacecraft carried augmented medical kits to deal with such events if necessary.

It is a tribute to the medical planning and execution of manned spaceflights that only one medical emergency, that involving Cosmonaut Vasyutin in late 1985, has been so extreme as to cause the abandonment of a mission. Many other minor medical problems have been recognized, diagnosed, and treated, however (see Chapter 2), and a medical kit appropriate to each mission's needs has been an essential part of logistic support from the earliest flights. The provision of drugs and equipment on board a spacecraft, and especially when no physician is in the crew, requires that the crew members be fully conversant with human physiology and anatomy under the conditions of stress met during spaceflight in both health and disease, and with necessary treatments. To that end, the training of all cosmonauts and astronauts has included comprehensive medical instruction, and this aspect will be discussed in the following chapter. Furthermore, both the contents of a medical kit and the training given to its potential users must rely in part upon predictions of the likely incidence of various medical events.

Both the Russians and the Americans have carried out extensive epidemiological surveys to assess the likelihood of such occurrences. These surveys have been based on large normal populations, on isolated populations in the Arctic, the Antarctic and on submarines, on laboratory isolation studies, and on previous experience in space itself. With time, of course, the validity of such predictions will be clarified but the kinds of figures estimated originally included a minor ailment every 1,500 man-days in space, a dental problem sufficiently serious to compromise a crewman's efficiency every 9,000 man-days, and a need for surgery every 28,500 man-days (or once every eight or nine years on an eight-man space station). And the need for evacuation was considered to be an extremely remote possibility at one event every 174,000 man-days! These estimates appear to be hopelessly optimistic in the light of experience so far, however, and will require considerable revision: minor ailments (including anorexia, backache, constipation, dermatitis, diarrhoea, headache, respiratory and urinary tract infections, fatigue, and insomnia) have been a feature of almost all missions, and the premature return to Earth of Cosmonaut Vasyutin for medical reasons occurred when the combined Russian and American experience in space had not even reached 6,000 man-days.

Russian spacecraft have carried comprehensive medical kits which contain a predictable assortment of analgesic (for pain), anti-emetic (for motion sickness), gastro-intestinal, anti-histamine, and sedative preparations usually in tablet form, although a few were injectable on command from the ground. In addition, drugs for use as protection against ionizing radiation have also been included, although they have not yet been used in flight since radiation levels have remained low. Non-specific tonics for fatigue, antibiotics, and general first aid equipment in the form of bandages, surgical tape, and haemostatic sponges complete the kit.

For the first four Mercury missions, an analgesic, an anti-emetic, and a vasoconstrictor (for shock) were the only drugs carried. All were in injectable form and could be self-administered through the spacesuit. For the final two Mercury flights, the vasoconstrictor was omitted but a stimulant (dextroamphetamine) was carried, and indeed was used to relieve fatigue prior to re-entry of Mercury 9. For the longer missions of Project Gemini, a more extensive medical kit was provided, with contents based largely on common sense. All of its drugs were in tablet and/or pre-loaded injectable form since other formulations, such as powders and glass ampoules, could clearly not be used conveniently in microgravity. In addition to the drugs carried previously, the Gemini kit included a selection of pain killers, decongestants (used to reduce the nasal stuffiness induced by microgravity, should a cold develop, and during descent to minimize the risk of barotrauma), an antibiotic (tetracycline), eyedrops (to ease irritation caused by the dry spacecraft atmosphere), and an anti-diarrhoeal preparation. Most of the drugs, and especially the stimulant, were used at some stage during the programme and, as with other projects, the response of each astronaut to the drugs carried had been assessed prior to flight to ensure that no allergic or untoward sensitivity reactions occurred.

For the Apollo and Skylab missions, some drugs were replaced by more effective preparations and yet more were added to the armamentarium. In particular, sedatives were included to aid sleep and, after the appearance of the cardiac abnormalities on Apollo 15, some anti-dysrhythmic drugs were included. Potassium supplements were also prescribed for flights following Apollo 15 in an attempt to counter the presumed mechanism of the dysrhythmia. The consistently most 'popular' drugs were pain-killers, sedatives, nasal decongestants, and anti-diarrhoeal tablets, while the others were used infrequently or not at all. Bandages, surgical tape, skin creams, and lip salves completed the kit; an abbreviated form of which made up the medical package carried on the lunar module. Additional equipment, including diagnostic aids and surgical instruments, was carried on board Skylab in view of the extended nature of its missions.

The Shuttle Orbiter Medical System (SOMS) includes a medicine and bandages kit containing a similar selection of items to its Apollo predecessor, an emergency medical kit, and various medical check lists. The Portable Oxygen System is also available for therapeutic use should the need arise, and future additions to the SOMS may include laboratory test kits for infections and a haematology test package. Finally, facilities also exist whereby a seriously ill or injured astronaut can be restrained within the cabin.

Post-flight medical care

Complete physical and laboratory examinations are conducted on all space travellers as soon as possible after return to Earth, and usually within the first few hours post-flight. When compared with the baseline, or control, levels pre-flight, many consistent changes have been noted in these examinations, and particularly in the cardio-vascular, musculo-skeletal, and neuro-vestibular systems. Further medical assessments are carried out for several weeks after spacecraft recovery in order to monitor any resolution of these changes: many do indeed return to pre-flight values, but some do not and, again, these findings will be discussed in detail in chapter 11 since they are of vital importance to the future of prolonged manned spaceflight.

Occasionally, it has been necessary to provide immediate medical treatment for returning crew members: for example, the lacerated scalp sustained by an astronaut during recovery of Apollo 12, the urinary tract infection in an Apollo 13 astronaut, the chemical pneumonitis in the Apollo participants of the ASTP, and more recently the mysterious illness of the Salyut 7 commander Vasyutin. By implication, therefore, physicians are always among the very first people to greet returning crews, and form a vital component of the recovery teams.

Lunar Quarantine Programme

One particular aspect of post-flight medical care was unique to the early lunar landings of Project Apollo: the Lunar Quarantine Programme (LQP). The specific aim of the programme was to ensure that the Earth and its environment were protected from any microbiological hazards associated with the return of lunar material. Secondary aims were the protection of lunar material and the operational integrity of the mission. It was assumed that, however unlikely, the Moon was indeed capable of supporting life and so a quarantine programme was justified. But, and this was crucial to the ability to define procedures, the risk was regarded as so small that quarantine would not take precedence over the preservation of human life. Studies of the invasiveness of disease-causing terrestrial organisms led to the conclusion that overt disease would be manifest within twenty-one days of exposure, and usually within a very much shorter period if the organism was particularly virulent. The crew quarantine period was therefore set at twenty-one days, beginning as soon as the astronauts left the lunar surface. Spacecraft housekeeping using special vacuum brushes and stowage bags minimized internal contamination during the return to Earth, and for the spacecraft recovery phase itself the astronauts wore biological isolation garments and a respirator to prevent any contact with their surroundings and the rescue personnel.

Once the astronauts had left the Command Module, its hatch was sealed and all external surfaces were decontaminated. The liferafts used by the crew and all decontamination equipment were then sunk at sea. The CM was returned to the recovery ship by helicopter where its hatch was attached to a tunnel from the Mobile Quarantine Facility (MQF), located on the hangar deck, through which a technician crawled to remove documents and samples from the cabin. The hatch was resealed and the CM transported to the Lunar Receiving Laboratory (LRL). Meanwhile, the astronauts, their equipment, and samples were isolated in the MQF. The facility could also be used on board an aircraft and while being transported by road, and so was inhabited until delivered to the LRL at the Johnson Space Centre.

The MQF could accommodate six people for ten days, and contained a lounge, a galley, toilets, and sleeping facilities. Basic medical equipment was also carried. The environment was maintained at a slightly negative pressure to ensure that no outbound leaks would occur, and all discharged air was carefully filtered. In addition, very stringent precautions were taken with all biological and inert samples. The crew members then spent the remainder of their twenty-one days' quarantine in the crew reception area of the LRL. The laboratory was a very large complex which also housed a sample operations area and an administrative support area. Incarcerated along with the astronauts were two flight surgeons, medical laboratory technicians, a recovery engineer, and cooks and stewards; all of whom had been carefully screened before the flight. For the astronauts, brief daily medical examinations were performed, and biological specimens obtained and analysed. And a fully comprehensive examination was carried out before release on the twenty-first day. Lunar samples also underwent exhaustive biological testing before being released from quarantine after fifty to eighty days.

Virtually the same LQP was imposed for the first three lunar landing missions: those of Apollo 11, 12, and 14. No extraterrestrial micro-organisms were isolated from any source, and release from quarantine of all human and inanimate components occurred on schedule. The programme was discontinued for subsequent flights, but has been described in some detail because a similar exercise will undoubtedly be required when future space voyagers return from visits to other worlds.

Longitudinal Health Studies

Finally, the detailed series of medical examinations performed throughout the careers of cosmonauts and astronauts provide an ideal data base for longitudinal health studies. As described earlier, the physiological data obtained from the pioneering programmes were, of necessity, of an observational nature and were based on the operational and safety requirements of the mission. With the advent of the Salyut and Skylab space stations, however, the opportunity to study a wider range of variables, from the purely physiological and medical standpoints, became a reality, albeit in a very small and highly selected group of individuals. Nevertheless, the data so obtained have greatly advanced medical knowledge, and reasonable predictions can now be made concerning a wide range of human responses to prolonged spaceflight and, just as importantly, to the return to normal gravity back on Earth (see Chapter 11). Of equal importance are the findings of cohort studies which may be related to the long-term health of whole populations. For example, an early study of the cardiovascular fitness of astronauts suggested, surprisingly, that the incidence of heart disease in the astronaut corps was not significantly different from its incidence in the general population. With the passage of time, however, the group has come to display the expected trend of a lower risk of heart disease. This particular finding is explicable in terms of the accepted higher standards of initial general fitness and continued medical care of this group, but other studies may give clues to the aetiology of maladies not as yet elucidated. With the large number of astronauts intended to fly on the STS programme, this kind of longitudinal study will have even more relevance since there will be no restrictions as to age or other arbitrary contra-indications for spaceflight: provided that the health of the individual, the health of others, and the success of the mission will not be jeopardized, permission to fly for legitimate reasons will be forthcoming. Short-term studies will provide additional data about the effects on individuals exposed to single missions. And indeed the motion sickness programme is one example of this kind of investigation. Medium-term studies will reveal the effects, if any, of repeated exposure to relatively short periods in space; and long-term studies, in the form of annual medical examinations, are envisaged, which will have occupational and environmental health importance for space flight.

Chapter ten

Selection and training

In his book *The Right Stuff*, Tom Wolfe endeavours to explain precisely what it is that makes a man wish to ride into space. He was, of course, dealing specifically with the 'Original Seven' Mercury astronauts, but exactly the same motivation must surely drive all the men and women, on both sides of the Iron Curtain, who have subsequently followed in the footsteps of the Seven and in those of their Russian peers. This chapter is not concerned with the ultimately indefinable qualities which prompt an individual to volunteer for dangerous enterprise, for that would involve a lengthy discussion of the philosophical concept embodied so well by Johannes Kepler in a letter to Galileo in 1610: 'Let us create vessels and sails adjusted to the heavenly ether, and there will be plenty of people unafraid of the empty wastes.' Rather, it deals with the situation once a man or woman has taken the decision to aim for the stars. It is therefore concerned with the selection criteria, with the exception of the medical aspects dealt with in the previous chapter, and with the subsequent training process required to make ready the willing volunteer for spaceflight itself.

Selection

To a large extent, the early cosmonauts and astronauts were a self-selected and pre-selected cadre since the majority were enlisted from the ranks of military test pilots, and those few that were not had been trained in the military. This is not at all surprising because it is intuitively clear that the temperamental characteristics which define test pilots (see previous chapter) would be expected to be the same as those needed for pioneering space travel. Furthermore, test pilots had already shown themselves able to tolerate the stresses of high performance military flying, and it was an entirely reasonable supposition that they would withstand the similar stresses of spaceflight; at least, they formed the most appropriate Earth-based model.

Russian selection procedures

Throughout their manned spaceflight programme, the Russians have adopted a policy of selecting and training groups of cosmonauts for each project. From the group, a sub-group was identified whose members all 'rehearse the flight dozens of times'. This concept of immediate flight readiness of several crews for each mission has stood the Russians in good stead, but is confusing for the outside observer. Furthermore, once the mission had been flown, the sub-group disbanded and its members were reallocated. The first cosmonauts were selected only from the Soviet Air Force, and the responsibility for selection and training fell to two senior Air Force officers, one of whom, Colonel Evgeni Karpov, was a specialist in aviation medicine.

The search for the Vostok crews began in October 1959, and six months later the first team of twenty cosmonauts began training. Twelve of this group ultimately went into space, including of course Gagarin. The remaining eight succumbed to ill health or were rejected for other reasons during the training phase. At the end of 1960, six men were chosen to form the immediate readiness sub-group and went into intensive training; but it was not until just four days before the historic flight that Gagarin was chosen to be the first man into space! The Vostok team was later supplemented by the inclusion of three women, including Tereshkova. By 1963, a further intake had entered training and included some military engineers as well as pilots. This group was more experienced and older than the original selection and, together with the remaining active cosmonauts from the Vostok programme, formed the nucleus of the crews for the two Voskhod missions. And so it continued throughout the Soyuz and Salyut eras: new cosmonauts were selected at regular intervals to begin training and to augment the teams of experienced pilots, engineers, and scientists which were always ready for flight. With the long-duration missions of the Salyut space stations, the selection process was enlarged to embrace both civilian engineers (often senior personnel responsible for the design of the craft they were destined to fly!) and foreign nationals, as part of the Interkosmos programme of international co-operation. The constitution of the team and its sub-groups was continually being altered, not least because the Russians tended not to allow their cosmonauts to undertake more than three or four spaceflights, although most 'retired' cosmonauts then entered an active role in the administration or training aspects of the space programme.

American selection procedures

NASA is a civilian organization and from its inception it was believed that its astronauts should be recruited from all available sources, including perhaps academic scientists, balloonists, deep sea divers, flight surgeons, mountaineers, submariners, and test pilots. But in 1958, President Eisenhower sensibly reversed the earlier policy and decreed that the initial intake of astronauts, for Project Mercury, should be drawn from the military test pilot population. Brief numerical details of the astronaut recruitment drives for this and subsequent projects were given in Chapter 9. But as well as the exacting medical standards required of a successful applicant for the American programme, several other essential criteria were imposed, and the general requirements defined for the 'Original Seven' included the following:

- American male,
- aged 25–40 years,
- less than 180 cm (71 in) tall,
- physically fit,
- graduate in science or engineering,
- 1,500 hours as a jet pilot,
- graduate of a test pilot school.

Clearly these requirements were incompatible with the aims of subsequent programmes, and many were relaxed for later intakes as the effects of spaceflight progressed from postulation to the knowledge of experience. Sex, age, and indeed nationality were no longer bars, and the educational and professional criteria were tailored to the needs of the mission. Throughout the history of American manned spaceflight, however, the selection of personnel has had as its foundation the five basic duties required of all astronauts and established by NASA's Space Task Group in 1958:

- to survive, i.e., to travel into space and return safely,
- to perform usefully in the space environment,
- to provide a back-up for automatic systems and so enhance reliability,
- to observe that which cannot be reported by instruments,
- to improve technology by acting as a true test pilot.

Training

The extensive training received by cosmonauts and astronauts in all aspects of space science (including astronautics, astrophysics, astronomy, celestial mechanics, computer science and electronics, flight skills, mechanical engineering, rocket design and propulsion, and so on) are outside the remit of this book. It is, however, concerned with those aspects of training which have a medical or survival implication; that is, with pre-flight physiological training, with medical training for use in-flight or post-flight, and with survival training. Similar methods of such biomedical training were developed by both Russian and American organizations.

Physiological training

The aim of physiological training is to acclimatize, and hopefully enhance, the

139

individual's response to the various physical and physiological stresses encountered during spaceflight (even though careful spacecraft design minimized the effects of these stresses). As such, the training for manned space travel was an extension of the proven procedures used in the world of conventional aviation medicine, and was largely based upon the precept that repeated exposure to a stress leads to its better understanding and a greater ability to cope.

Russian physiological training The training programme for cosmonauts entailed a progression through three main phases of activity, and could last up to two years. During the first phase, use was made of general techniques to improve physical fitness, based on the reasonable assumption that improved fitness would also help the ability of the cosmonaut to withstand the various stresses of spaceflight. Of particular concern was the behaviour of the vestibular apparatus when in space, problems with which had been predicted for several years prior to manned spaceflight by both the Russians and the Americans. Both nations had employed aircraft flying Keplerian trajectories to study the effects of short-term microgravity. (Keplerian trajectories are aircraft flight paths which describe a parabola. During the ten to fifty seconds or so at the height of the parabola, the centrifugal force developed by the aircraft exactly matches the centripetal force of Earth's gravity attracting it, and a condition of microgravity exists).

The results of these flights in the late 1950s suggested that, while some subjects could tolerate microgravity with no ill effects, many found the experience discomforting because of the illusory sensations it induced, and others were extremely distressed. Fortunately, some resistance to these changes could be acquired by repeating the flights, although for some non-professional aircrew this could mean thirty or more rides in what has come to be known as the 'Vomit Comet'! Professional aircrew tended to tolerate the periods of microgravity well, or to require only ten to twelve repetitions. Keplerian parabolas therefore became part of the training and conditioning process used by both the USSR and the USA for their space crews. During these flights, the cosmonauts and astronauts experienced microgravity while restrained and when untethered so that they could float about the cabin. More recently, the value of such procedures has been extended to provide testing of equipment under conditions of microgravity and to allow practice of manual tasks.

Other ground-based techniques, including rotating and swinging devices have been employed to reinforce the vestibular conditioning. Exposure to progressively increased transverse (Gx) accelerations on a centrifuge was adopted to increase tolerance to the accelerations of launch and recovery. Cosmonauts whose original tolerance was relatively poor have particularly benefited by a gradual increase in acceleration exposures to 10Gx. Education and training in active methods of increasing tolerance, such as muscle tensing and correct breathing patterns, accompanied the centrifuge programme. Since one possible emergency scenario was a failure of spacecraft heat dissipation, cosmonauts have also been exposed to a thermal challenge, while undressed, at 60°C; while hypobaric (decompression) chambers were used to demonstrate the effects of hypoxia and decompression. The general physical fitness programme augmented all of these training procedures. It involved both group activities in the gymnasium and individual work on treadmills, ergometers, and trampolines, as well as personally preferred sports such as swimming, jogging, and water skiing.

The second phase of cosmonaut training, which in fact progressed in parallel with the first, was the use of simulators to increase the operational skills required for flight: including those of flying itself, navigation, communication, and approach, docking, and landing. And, predictably, the third phase attempted to combine the first two in real-time simulations of the entire planned mission carried out in a realistic Earth-based spacecraft model. Part of the preliminary training for this final phase involved several days in a sound-proof chamber during which work-rest patterns were altered to mimic those of a space flight and the behavioural responses of the crew were studied intensively.

The training of scientist-cosmonauts differed little from that described for their pilot colleagues, although much of the first phase, and especially the implementation of vestibular conditioning, was carried out while the individual was still at his primary place of work. In addition, acclimatization at high altitudes in the mountains was instituted to improve tolerance to hypoxia and, apparently, to increased accelerations.

American physiological training Not surprisingly, the American astronaut training programmes were very similar to those of their Russian counterparts. Again, great emphasis was placed upon improving and maintaining general physical fitness; and simulation of, and conditioning to, the stresses of space-flight were vigorously pursued. As an example, the Mercury astronauts endured four centrifuge training programmes to enhance their abilities to control the craft under conditions of increased accelerations during launch and re-entry. The first two programmes were for engineering and familiarization purposes, and the others for intensive operational preparation. The centrifuge gondola (cabin) was capable of being filled with oxygen and decompressed, while computer control allowed the device to follow either a ballistic or an orbital mission acceleration profile. Some runs were conducted at simulated altitudes of 28,000 feet (8,536 m) with the subjects wearing full pressure suits.

Other programmes involved exposure to altitude in decompression chambers, to heat, noise and vibration, to isolation, and to simulated microgravity. And, a specially designed and constructed piece of apparatus, the Multiple Axis Space Test Inertia Facility (MASTIF) which could simulate violent tumbling of a spacecraft, was used to train the astronauts in recovery techniques therefrom. This device was particularly Draconian and was not used for subsequent projects!

Indeed, for the Gemini, Apollo, and Skylab astronauts, the emphasis in training switched, in the light of experience, from the intense preparation for the physical hazards of spaceflight to a more considered approach combined with the complex operational needs of, for example, rendezvous and docking, free-space EVA, and lunar-surface activity. Furthermore, the crews now included the scientist-astronaut element, several of whom were additionally trained as jet pilots. Although physiological training continued (the heat tests were omitted, however), the use of sophisticated and high fidelity spacecraft simulators evolved and increased with the complexity of the programme, and all crews spent many hours in the mission simulators at the MSC. Real-time simulation of routine and emergency situations was found to be crucial to the success of each programme, and the astronauts were trained, in many such procedures, almost to the point of being automatons. Much cross-training was necessary in that, as well as their primary tasks, each crew member had to be able to control the spacecraft propulsion systems and, most importantly, to be able to accomplish re-entry and landing manoeuvres.

A few aspects of flight preparation were found to be somewhat irksome, however, and this was especially true of some of the baseline studies for in-flight medical experiments. After one such experience before the flight of Gemini 7, attempts were made to reduce the workload on crews in the immediate period before launch. Of course, training procedures were continually updated in the light of flight experience, and one example of this was the increased attention to simulated microgravity training for EVA following the exhaustion of the Gemini 9 and 11 astronauts (see Chapter 8). After the Gemini 9 mission, a neutral buoyancy water tank, the Weightlessness Environment Training Facility, was used to assess the adequacy of the additional restraint devices added to the spacecraft exterior (Plate 12). The simulation was considered to be so realistic of microgravity, however, that formal astronaut training in the tank was implemented for all missions from Gemini 12 onwards. The technique not only established the feasibility of performing individual tasks during EVA, but also enabled the astronauts to learn how to relax by minimizing the activity of muscle groups not required for specific manoeuvres.

Finally, although the general training for all astronauts was similar, it was a policy that each member of a project group was given specialist areas of responsibility in which to become expert. Thus, for example, for Project Mercury, Shepard studied the recovery systems in detail, Glenn the capsule lay-out, Schirra the life support systems, and so on. For Project Gemini, John Young, now the most experienced American in space and the Senior Astronaut, was the expert in environmental systems and survival equipment.

Medical training

As implied in the previous chapter, the provision of extensive medical supplies, even when combined with advice from ground-based physicians, is of little benefit unless there are people on board the spacecraft capable of timely and appropriate action in their use. Furthermore, since physicians are not immune from illness or injury, there is a clear need for some medical training of all cosmonauts and astronauts, whether or not a doctor is in the crew.

Cosmonauts receive sufficient medical training before flight to enable them to render aid to others in the crew and, if necessary, to themselves. Extensive briefings on anatomy, physiology, likely clinical illnesses and their differential diagnoses are given, as well as detailed information on treatment and the effects of any drugs carried. In the United States, too, general medical training is provided, along similar lines, as a sixteen-hour course. This course covers particularly the anatomy and physiology of the cardiovascular system, the pulmonary system, the neuro-vestibular system and the visual apparatus; and their responses to both increased and decreased (microgravity) accelerations. Methods of physiological investigation of these systems and the significance of possible findings are also discussed, as are the laboratory results of previous programmes. The psychological aspects of crew selection, confinement and isolation, closed group interactions, and the response to stress complete the curriculum.

Because of the prolonged nature of their missions, the Skylab astronauts were even more intensively trained, and an additional three-day course was given. This covered instruction in diseases of the cardiovascular, pulmonary, abdominal, and musculo-skeletal systems, as well as of the head and neck, and skin. Dental equipment and minor oral surgery were also demonstrated (Plate 9). Medical checklists were provided to give in-flight guidance, and crews were able to carry out practical procedures such as bandaging and splinting, nasogastric intubation, catheterization of the bladder, tooth extraction, and even tracheostomy. It was, however, intended that the more complex procedures would be under the direct supervision of ground-based clinicians! Astronauts are also required to undergo a three-year refresher course in both the theoretical and practical aspects of space physiology and medicine. And a mission-specific medical briefing is given about one month before a flight during which much detailed information is offered in many areas including:

- pre- and post-flight medical procedures, including details of the health care programme, and emergency medical support,
- composition and usage of in-flight medical stores (drugs and equipment),
- provision of food and water,
- hygiene and the waste management system,
- use of any medical instrumentation (operational and experimental), including radiation monitors, and logging of data.

Table 10.1 Medical training for Shuttle crews

General	Location, organization, and use of medical equipment, drugs, and checklists
Vital signs	Pulse, blood pressure, temperature, respiratory frequency, pupil size, and reaction
Physical examination and treatment	*Eye* Ophthalmoscopy, lid eversion, staining, removal of foreign body
	Ear Auroscopy
	Nose Control of epistaxis (nosebleed)
	Throat Insertion of airway
Control of haemorrhage	Direct pressure, pressure points, tourniquets, use of Steristrips
Bandaging	Wounds of extremities, chest, and abdomen
Splinting	Neck, fingers, and extremities
Dental treatment	Temporary fillings, gingival injections
Emergency procedures	One-man cardio-pulmonary resuscitation, abdominal thrust, tracheostomy
+Monitoring equipment	Radiation monitors, operational bioinstrumentation, including that for EVA, and ECG

As an example of the thoroughness of the medical training offered, Table 10.1 lists that undertaken by Shuttle crew members.

Survival Training

A necessary part of manned spaceflight operations is the provision of an emergency medical support service during launch and landing. An efficient rescue organization requires trained and practised personnel with a rapid means of transport to the incident site, and equally rapid access to full medical facilities subsequently if necessary (Plate 10). The entire team responsible for this service, including the spacecraft crews, is trained in its implementation; and, as with other forms of training, procedures have evolved with experience. For example, the near drowning of Astronaut Grissom in the sight of his rescuers on his return to Earth aboard Mercury 4, led to subsequent crews wearing a sealed neck dam on their spacesuits to prevent water entering.

Fortunately, most spacecraft recoveries so far have been without such dramatic consequences, and have been within the planned rescue zone (an exception was the aborted mission of Soyuz 18A, which landed near the Chinese border (see Chapter 6)). But this may not always be the case and crews must be prepared to survive, perhaps for many days, in potentially harmful environments. Survival training is therefore an integral part of flight preparation: even conventional water or land recoveries may occur many miles from the rescue forces and, of course, Russian crews must be ready for water landings, while American crews could recover over land. Thus, any returning space traveller may unexpectedly find himself isolated in extremes of temperature at sea, on land, in a desert, or up a mountain. Although primarily for in-flight situations, the comprehensive medical training given to spacecraft crews is clearly of value following such a return to Earth in a remote location which may threaten survival. But specific survival training must also be given. Training revolves around the need to survive in cold, dry situations (the Arctic), cold, wet situations (the sea), hot, dry situations (desert), and hot, wet, or moist situations (tropical seas or jungle). The major physiological problems associated with these conditions have been discussed earlier; the direct effects of temperature extremes in Chapter 6, and the effects of dehydration, and the necessity for adequate emergency rations of food and water, in Chapter 7. There are some other hazards, however, such as frostbite and snow-blindness in the cold, sunburn in the desert, infections in the moist heat of jungle, and even sharks in the tropical oceans. All of these issues are presented to cosmonauts and astronauts, and some practical training in desert and water survival techniques is given.

Part three

Keeping Man in Space

It is quite apparent that modern science and technology have been able to place men in space and, with a few tragic exceptions, to return them safely to Earth. The lessons of the manned spaceflight programmes, however, tell us that if we are to stay in space for any length of time – that is, beyond the current duration record of nearly eleven months – then other aspects of the hostile environment must be considered, and either controlled or conquered. The most pervasive of these aspects is the effect of microgravity upon human physiology both while in space and on return to Earth. Although this major topic has been touched upon in Parts 1 and 2, the first chapter in this part deals exclusively with this unique threat and its impact on various body systems. Not surprisingly, the mind too may be affected by spaceflight, and the first section of the second chapter discusses the undoubted psychological import of prolonged space travel, with its implications for selection and training, confinement and isolation, disruption of circadian rhythms, and inter-personal and inter-sexual relationships. A second section addresses the important physiological and psychological points raised by the inclusion of women in space crews. Some aspects of spacecraft habitability not yet considered, such as man-machine interaction and facilities for entertainment, relaxation, rest, and privacy, form a third section. The final chapter discusses the future of man's adventure into space, with particular emphasis on the likely progress needed in the areas covered by earlier chapters. Most immediately relevant from the aspect of survival is the need for advanced, self-contained life support systems and some form of artificial gravity, without which long journeys into the unknown cannot even be contemplated.

Chapter eleven

Decreased accelerations: microgravity

It will be recalled from Chapter 1 that the Earth's gravitational field is still apparent at a distance of several million miles into space. Thus, only when entirely free of gravity, whether it be from our own gravitational field or that of another planet, can a condition of true weightlessness be said to exist.

Furthermore, even tiny perturbations in a spacecraft's environment, such as the firing of attitude control rockets or the movement of a crew member, can induce small changes in the gravitational field: once in Earth orbit, or having achieved Earth escape, spacecraft and their occupants are exposed from 1×10^{-4} to 1×10^{-5}G, with increases to about 1×10^{-3}G during some in-flight manoeuvres. Because of this finite level of force, the more appropriate and accurate term *'microgravity'* has been used throughout this book. The conditions within spacecraft, however, are so close to zero gravity, when compared with the usual experience of the 1G environment of Earth, that the distinction is really academic as far as physiological observations are concerned. (The conditions are somewhat more critical when the need to study and manufacture pure materials is space is considered: hence the interest shown by materials scientists in free-flying, unmanned laboratories.)

Naturally, microgravity affects all body systems to some degree and many studies of human physiology, before, during, and after flight, have been accomplished throughout the manned spaceflight programmes (Plate 13). Of particular concern to the future of man in space are changes affecting the neuro-vestibular system, the cardiovascular system, fluid and electrolyte balance, haematology the musculo-skeletal system, body mass, and exercise capacity. This chapter describes the changes observed in these systems, and in others of apparently less serious import, as well as the effect of various countermeasures. The results of these studies, however, must be viewed with circumspection since the number of subjects involved has been necessarily limited, the subjects themselves have been highly trained and very fit, and the extensive use of any remotely beneficial countermeasure has often been imposed in order to reduce detrimental effects. In addition, and especially in the early programmes, study of the physiological responses to the space environment was very much subservient to the operational needs of the mission. In later flights, however, and particularly those of Skylab, Salyut, and some Shuttle missions, physiology has been studied for its own sake. And, despite all these experimental limitations, a vast amount of valuable information has now been collected concerning the response of men, and women, to the unique environment of microgravity.

The neuro-vestibular system

The neuro-vestibular system is that part of the central nervous system which is specifically responsible for the perception of the body's spatial orientation and for the maintenance of its equilibrium. These functions are accomplished by a complex interaction between information supplied to the brain from the eyes, the inner ears, and the proprioceptors (position sensors) in the muscles, joints, and tendons. All of these inputs are combined to provide a knowledge of the body's position, attitude, and movement in relation to fixed references which, on Earth, are usually the force of gravity and a horizon. From the earliest age, these powerful external influences combine to provide perception of correct orientation. So strong is this conditioning that movement in an abnormal manner, as for example on a fairground swing, in a boat, or in a military fast jet aircraft, may so disrupt the brain's normal interpretive mechanism that illusions of movement may occur, or motion sickness may develop. It is with the latter that space travel has become particularly associated.

Of the three sensory modalities concerned with orientation, that of sight is the most important in virtually all circumstances for, if visual function is normal and external visual cues are unambiguous, the eyes provide very reliable sensory information. The same can usually be said of the organs of balance which make up the inner ears: the semicircular canals and the otolith organs (Figure 11.1.).

The semicircular canals, three on each side aligned mutually at right angles, provide information concerning the angular (three-dimensional) motion of the head. They may be regarded as angular (radial) accelerometers, and the two sides work as matched pairs to signal to the brain the direction and magnitude of movement in pitch, roll, and yaw. Each canal is filled with fluid called endolymph, the inertia of which causes its movement within the canal in the opposite direction to that of the head in that plane. The endolymph passes over, and bends, the hairs of sensory cells (collectively termed the crista) embedded in the wall of the canal at one end (the ampulla). A mass of gelatinous material, the cupula, sits on top of the hairs, and it is the displacement of this cupula by the moving endolymph which distorts the hair cells and causes them to send impulses to the brain via the vestibular division of the 8th cranial nerve. In the laboratory, sustained *changes* in the velocity of the head greater than about $3°.sec^{-1}$ will be detected by those crista in the

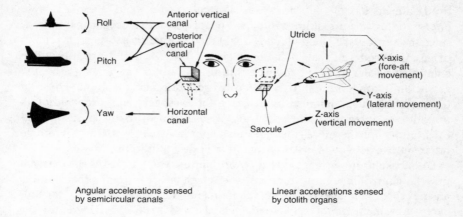

Angular accelerations sensed
by semicircular canals

Linear accelerations sensed
by otolith organs

Figure 11.1 Relationship of the various components of the vestibular apparatus to their axes of response, and to the planes of spacecraft
movement

canals lying in the plane of movement. Because they are accelerometers,
however, the signal dies away when a constant velocity is reached (that is,
when there is no longer any acceleration) even though movement is
continuing. Similarly, a deceleration will be detected by the appropriate
canals but, again by virtue of its inertia, movement of the endolymph will
continue for a short while even after the head is stationary. In the absence of
corroborating visual signals, a false sensation of continued rotary motion will
persist.

The otolith organs may be regarded as linear accelerometers, and are found
within the saccule and utricle of each inner ear in close approximation to the
semicircular canals. The saccules and utricles are small sac-like dilatations in
the structure of the inner ear, lie in the vertical (z) and horizontal (x and y)
axes respectively, and respond to accelerations in those planes. Groups of hair
cells, this time called maculae, again comprise the signalling mechanism; their
distortion being achieved by the inertial movement of a membrane which
overlies them, and in which is embedded small crystals (otoliths) of calcium
carbonate. Linear accelerations greater than $0.1m.sec^{-1}$ may be detected
under controlled conditions, but the signals decay once constant velocity has
been achieved. (As with signals from the semicircular canals, however,
thresholds of detection are likely to be considerably higher once outside the
laboratory and subject to all the sensory inputs of normal life.) The otolith

organs are, most importantly, responsible for the appreciation of motion in relation to Earth's gravity, but they too can be misled once movement in the third dimension of air or space is undertaken, and their accurate interpretation again requires visual and proprioceptive mechanisms. Indeed, since the vestibular system alone is unable to distinguish between movement of just the head and that of the entire body, the visual and proprioceptive components and their integration within the brain (particularly by the cerebellum) are essential to correct neuro-vestibular function. Thus, in terrestrial life, the combined inputs from the eyes, ears, and proprioceptors provide an integrated and accurate assessment of body position and orientation. Furthermore, the vestibular apparatus also acts to co-ordinate reflex muscle activity in order to maintain body equilibrium.

Of particular importance is that co-ordination which involves the external muscles controlling eye movements, and which under normal circumstances ensures that the image focused on the retinae of the eyes is static. To this end, every movement of the semicircular canals, and to a lesser degree of the otoliths, is matched by compensating movements of the eyeball so that even the slightest head movement is countered by an equal but opposite eye movement, so ensuring that the retinal image remains undistorted.

All of these mechanisms have evolved for life on Earth, and the dynamic responses of the organs concerned remain appropriate for the relatively sedate motion of man on foot. Once mechanical transport became available, and especially when it involved the third dimension, the adaptive abilities of visual, vestibular, and proprioceptive mechanisms could easily be swamped, and the interpretive ability of the brain overcome. This is very important in the world of conventional military and civilian aviation, when many incidents and accidents have been attributed to false impressions or illusions of angular and linear motion brought about by increased stimulation of the sensory modalities: that is, to disorientation. But, although intensely increased accelerations, especially linear, are a feature of spaceflight (Chapter 4), these are at entirely predictable phases of a mission, and are unlikely to cause significant problems of disorientation. Vestibular dysfunction in the form of space motion sickness, however, has been and remains a major concern.

Space motion sickness

The term sickness is, technically, somewhat of a misnomer since the syndrome is a manifestation of the normal physiological response to real or apparent, but unfamiliar, motion stimuli. There can, however, be no doubting the misery it engenders! The features of overt motion sickness are, unfortunately, familiar to a fair proportion of people 'enjoying' the benefits of modern forms of transport. General feelings of stomach awareness and epigastric (upper abdominal) discomfort are followed by increasing nausea, pallor, and sweating. There may then be a rapid worsening of the condition (the so-called avalanche phenomenon) with increased feelings of warmth, intense perspiration and salivation, and of course vomiting, which provides rapid, if temporary, relief of symptoms. Other features seen or felt may include widely dilated pupils, frontal headache, flatulence, hyperventilation, anxiety, apathy, depression, fatigue, and drowsiness.

Of course, the incidence and severity will depend upon the particular environment involved: for example, virtually all those exposed to very heavy sea states in small boats are affected to the point of vomiting, while in commercial airline operations only about 1 per cent are regularly stricken, and only a further 7 per cent in the worst air turbulence. In space, about 30 per cent of all cosmonauts and astronauts have been affected to some degree, while the incidence among crews of larger spacecraft is approaching 50 per cent. The second Russian into space, Cosmonaut Titov, was the first to report being motion sick in August 1961, and it has been a regular feature of Russian programmes ever since. But it was not until the early Apollo missions that American astronauts first reported a serious problem with what has now become known as one manifestation of the Space Adaptation Syndrome. On some flights, the condition has been so severe as to threaten the success of certain operational activities, and occasionally even that of an entire mission.

The aetiology of motion sickness is not entirely clear, although it has been the subject of debate since ancient times. One widely held modern view is that propounded by the British physiologist Reason, and termed the Sensory or Neural Mismatch Theory. This suggests that motion sickness develops whenever the sensory information provided by the visual, vestibular, and proprioceptive apparatus is at variance with that expected on the basis of previous experience. This pragmatic theory is able to account for most circumstance in which motion sickness occurs. Thus, for example, when sea sick below deck, strong vestibular inputs as a result of wave movement are not accompanied by the appropriate visual information. Similarly, simulator motion sickness is explicable because the visual information presented by static stimulators is not accompanied by appropriate movement and hence vestibular signals. In space, while visual and semicircular canal functions are retained much as on Earth, the virtual absence of gravity renders the otoliths impotent: the maculae depend for their stimulation upon the weight of the otoliths but cannot function as there is no weight! A direct relationship between rapid head movements and the occurrence of space motion sickness was noted by Titov, and this certainly seems to have a bearing on the in-flight occurrence. Table 11.1 lists the incidence of motion sickness in the various manned spaceflight programmes, and the low incidence during the Mercury and Gemini programmes, in which head movements were restricted by body restraint and lack of room, supports this notion.

Despite the attractiveness of the mismatch theory, it cannot explain why motion sickness takes the form it does, or indeed why it should occur at all. But it does suggest the means by which the observed adaptation to a

Table 11.1 Incidence of motion sickness in the manned spaceflight programmes (to end 1982)

USSR			USA		
Programme	Crew	Affected	Programme	Crew	Affected
Vostok	6	1	Mercury	6	0
Voskhod	5	3	Gemini	20	0
ASTP	2	2	ASTP	3	0
Soyuz/Salyut	71	35	Apollo	33	12
			Skylab	9	5
			Shuttle	12	5
TOTAL	84	41 (50%)		83	21 (25%)

Note the increased incidence of 45 per cent in the larger spacecraft (Soyuz/Salyut, Apollo, Skylab, and Shuttle) compared with 11 per cent in the smaller craft, and an overall incidence of 37 per cent.

prolonged change in environmental motion takes place. Furthermore, it predicts that, once adaptation has occurred, a return to the original environment, whether it be a return to shore for a sailor or a return to Earth for a space voyager, may be accompanied by a recurrence of motion sickness while readaptation takes place. Vestibular adaptation to microgravity typically takes two to four days, after which head movements may be made with impunity for the rest of the time in space; but some returning crew members have indeed developed further motion sickness on re-exposure to the 1G environment. Other proposals have attempted to explain certain features of space motion sickness and, although some complement the mismatch theory, none is entirely satisfactory. One such complementary idea suggests that minor differences in the weight of the otolith organs in each ear require central (brain-based) compensation which is built up from birth, only to be de-stabilised on exposure to microgravity and again on return to Earth. Another, as yet unsubstantiated, concept suggests that the redistribution of body fluids, with the consequent engorgement of the head and neck may somehow influence the behaviour of the vestibular apparatus.

Space motion sickness typically begins to develop in those prone to it soon after entering orbit, and hence at a critical stage in most missions. Because of this, it has been said that the condition 'represents the greatest research challenge facing life scientists in contemporary space medicine and physiology'. Although this may be overstating the case somewhat, a very large effort has been made to elucidate the condition, to identify susceptible individuals, and to prevent its development by psychological, pharmacological, and mechanical means. As described above, however, despite the devoted attention of many scientists involved with the space programme, the precise physiology of motion sickness remains ill understood. On the ground, many provocative techniques have been employed to assess the susceptibility of prospective space travellers; including vertical oscillators, swings, roll and pitch rocking devices, visual stimulation, and, in the air, parabolic flights. Head movements during such manoeuvres are particularly nauseating and have proved very useful in predicting susceptibility to 1G motion sickness; but not to that in microgravity! Nor does a past history of Earth-bound motion sickness seem to have any predictive use. Reliance has, therefore, been laid on prophylactic methods to prevent or minimize the condition in space. Training before spaceflight has been attempted and appears to induce some reduction in susceptibility: rotating chairs, aerobatic and parabolic flights have all been used. Of course, the effectiveness of such measures is virtually impossible to assess because use is made of other in-flight protective methods. Psychological training in biofeedback techniques has proved valuable in returning conventional pilots, grounded because of motion sickness, to flying duties; but the role of this in spaceflight has yet to be established.

The use of anti-emetic (anti-sickness) drugs, however, has been highly successful, as might be predicted from their beneficial use on the ground. Many such drugs are available but virtually all have some side effects, such as blurring of vision, dryness of the mouth, and, significantly, drowsiness. Consequently, cosmonauts and astronauts evaluate drugs of choice before flight for both potential side effects and vestibular effectiveness. For astronauts, all those flying in space for the first time are treated prophylactically, as are those who have experienced motion sickness on previous flights; although such individuals seem less prone to sickness on subsequent missions, suggesting some form of permanent adaptation. An early and effective drug combination used in space was of promethazine hydrochloride and ephedrine sulphate, but the drugs most commonly used now in the STS programme are an oral combination of scopolamine and dextroamphetamine. Scopolamine can also be delivered through a patch of drug-impregnated material stuck to the skin. This transdermal technique allows a longer period of drug action than the oral route and, although early formulations were ineffective and produced marked side effects, recent trials have been most promising.

Lastly, since body movements, and especially those of the head and neck, exacerbate space motion sickness (as indeed they do for the conventional Earth-based syndrome), the Russians have employed a restraint system to restrict the motion of the head. The device, termed a Neck Pneumatic Shock-Absorber, consists of a cloth cap to which rubber cords are attached via the shoulders from a waist belt. The strength of the cords is such that considerable effort is required to move the head from an upright and forward-facing position. The Salyut 6 cosmonauts used the device with some success in reducing the severity of symptoms during the early days of their missions.

The title Space Adaptation Syndrome is most usually applied to the phenomenon of space motion sickness and adaptation thereto, but other manifestations attributable to the neuro-vestibular system have also been described. Thus, postural illusions, dizziness, vertigo, and nystagmus (rapidly alternating eye movements) have all occurred as immediate reflex responses to the environment of microgravity. Some of these, for example, dizziness, may precede the development of frank motion sickness; and indeed this was the classic progression of events related by Titov. Many cosmonauts and astronauts have reported illusions of postural type, and especially of apparent visual displacement of objects within the spacecraft or of hanging upside down. Such illusions are usually manifest immediately after entering microgravity and may last for several hours. They have also tended to recur throughout flight, especially at times of increased motor or visual activity. Disorders or equilibrium and gross muscle function, often manifest as an ataxia (lack of co-ordination) with an unsteady and wide-based gait, may also be a feature immediately after return to Earth, particularly when assessments are made with the eyes shut. Recent Russian evidence suggests that recovery may take many days or even months after very long missions, and that the role and importance of the proprioceptors in this post-flight phenomenon may have been seriously underestimated.

The cardiovascular system

The principal functions of the cardiovascular system are the delivery of appropriate amounts of oxygen and other nutrients to the tissues, and the removal of metabolic waste products therefrom. The heart is the central component of the system, while the blood vessels are the peripheral elements. Figure 11.2 shows a simplified version of the cardiovascular system. The heart may be regarded as two independent but matched pumps, each of two chambers (one atrium and one ventricle), working in series. The high-pressure, left side of the heart receives oxygenated blood from the lungs and pumps it to the delivery vessels (arteries and arterioles) of the systemic (i.e., non-pulmonary) circulation, and thence to the small capillary exchange vessels in all tissue beds. Within the tissues, oxygen and other pabula are given up by the capillary blood, and waste products (metabolites) are collected for onward transmission in the venules and veins, and thence back to the heart. Having traversed the low-pressure, right side of the heart, the blood returns to the lungs where carbon dioxide is eliminated and oxygenation recurs. Other metabolites are dealt with by the liver and kidneys during subsequent circulations, just as absorbed foodstuffs from the gut are picked up during one pass and delivered to the tissues later.

For a normal individual under resting conditions, the heart beats at a rate of seventy to eighty per minute, and each beat delivers about 70 ml of blood (the stroke volume) to both the systemic and pulmonary circulations; the cardiac output is therefore approximately 5.5 L.min^{-1} (9.7 pints.min^{-1}), rising

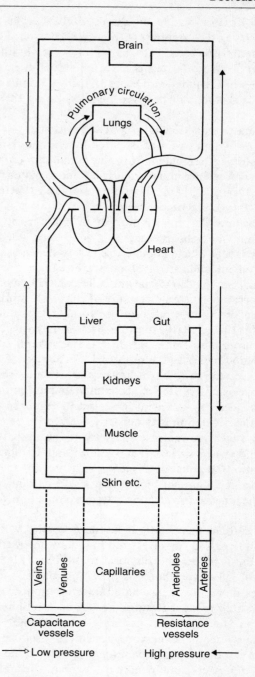

Figure 11.2 The cardiovascular system

to perhaps 25 L.min^{-1} during strenuous exercise. The magnitude and distribution of the cardiac output are, however, controlled according to the needs of individual tissue beds. Thus, blood flow to the brain is maintained at a constant level at all times, but that to individual muscle beds will vary according to the work required and being performed. Similarly, blood flow to the gastrointestinal tract will be given priority during the digestive period after a meal.

The other major function of the cardiovascular system is that of temperature control, and variation of blood flow to the skin is the means by which this is achieved (see chapter 6). The pulmonary circulation acts as a very efficient reservoir since it must clearly be able to match any alterations in systemic needs. The lungs are, therefore, unique in that they alone of all tissues must be able to accept the entire cardiac output at all times.

In order to achieve its primary purpose of maintaining tissue perfusion, adequate pressure must be present in the delivery side of the systemic circulation. The rate at which blood is able to flow depends upon the presence of a pressure gradient, and arterial blood pressure is itself a function of blood flow and resistance to that flow. The muscular contraction of the left ventricle of the heart generates an arterial pressure of about 120 mmHg (16 kPa) at its peak (the systolic pressure), while pressure during relaxation is maintained at about 80 mmHg (10.7 kPa) (the diastolic pressure) by virtue of the resistance residing in the arterioles. Sphincters in these peripheral vessels control the rate at which blood passes to the rest of the circulation. If the arterioles are the resistance vessels of the cardiovascular system, the veins are the capacitance vessels and are able to accommodate the run off from the tissue beds even when delivery is vastly increased in response to demand. The venous side of the circulation is at a progressively lower pressure until, by the time blood leaves the right ventricle on its way to the lungs, the pressure is just 25 mmHg (3.3 kPa) in systole (contraction) and 10 mmHg (1.3 kPa) in diastole (relaxation). Because of this lower pressure, and the large capacity of the lungs, the same volume of blood (that is, the cardiac output) is able to pass through the lungs as through the systemic circulation, and over the same period of time.

Control of cardiac behaviour and indeed of the whole circulation is extremely complex and involves the intimate interaction of intrinsic, autonomic, central, reflex, and humoral (fluid (blood) borne) mechanisms. Intrinsic control or autoregulation is the ability of the micro-circulation to vary its own resistance, and hence blood flow, in response to local variations in blood pressure, tissue pressure, and concentrations of oxygen and metabolites. Autonomic control describes the automatic neural influences exerted by the sympathetic nervous system to produce vasoconstriction (narrowing) or vasodilatation (widening) of peripheral blood vessels, and so to control blood pressure; and by both the sympathetic and parasympathetic nervous systems (see also 'Central and Peripheral Nervous System', p. 175) to influence the behaviour of the heart. Not surprisingly, the brain itself may influence central control over the circulation, both at an automatic level from specialized centres in the brain stem and mid-brain, and at a more conscious level from the cortex. Yet more mechanisms are reflex in nature and produce immediate responses as a result of local blood pressure changes (baroreceptors), and of changes in the chemical composition of the blood with respect to oxygen, carbon dioxide, and acidity (chemoreceptors). And finally, other body substances, termed humoral agents, such as cortisol and histamine, may exert their own influence.

On Earth, all of these controlling mechanisms combine to allow an accurate and adequate circulation to be delivered to all parts of the body under all normal conditions. Implicit in this ability is the need to be able to function when changing from the horizontal position of rest to the vertical position of most human activity. Specifically, as a result of evolutionary adaptation to the upright posture, the cardiovascular system of a healthy individual is well able to cope with the additional stress of hydrostatic pressure gradients imposed whenever the body's z axis is perpendicularly aligned to the force of gravity (see chapter 4). If the cardiovascular controlling mechanisms were not able to respond to this re-orientation, about 500 ml of blood would pool in the distensible veins of the lower limbs under the influence of gravity. In addition, the increase in hydrostatic pressure beneath the level of the heart leads to a further loss of fluid by transudation from the capillaries. If allowed to proceed unchecked, the loss of effective circulating blood volume would produce a fall in cardiac output and blood pressure, and cerebral blood flow would diminish. Therefore, whenever the upright posture was adopted, the individual would simply faint as a result of this orthostatic intolerance. Some pathological conditions in which the cardiovascular compensatory mechanisms have been damaged produce just such a phenomenon; and so, too, does exposure to microgravity. Thus, just as the imposition of *increased* accelerations of launch and re-entry outstrips the ability of the cardiovascular system to compensate (see Chapter 4), so the imposition of *decreased* accelerations, in the form of microgravity, causes the controlling mechanisms to re-adapt to the new environment.

Predictably, problems arise on return to Earth when a 1G environment is re-encountered. Orthostatic intolerance is regarded as potentially one of the most serious cardiovascular consequences of manned spaceflight.

Post-flight orthostatic intolerance has been a constant feature throughout man's experience in space, and is directly attributable to the abolition of all hydrostatic pressure gradients in the environment of microgravity. The absence of these gradients causes a major shift of fluid from the lower body towards the chest and head during the first few days in space. The rate of shift, as assessed by repeated measurements of thigh and calf circumferences, reaches a peak after about twenty-four hours, and the new steady state is attained within three or four days. Limb measurements show a rapid decrease

159

in circumference (of up to 30 per cent) over this time, to a level which is then sustained for the rest of the flight, although some cyclical variation has been observed in the very long Salyut missions.

In parallel with this decline in limb size, there is an objective and subjective increase in fluid present in the head and neck. Thus, photographs of the Skylab crewmen taken during the flight show a general puffiness (oedema) of the face, and particularly of the tissues around the eyes and the eyelids themselves. And the veins of the forehead, scalp, and temples, and of the neck, including the large jugular vessels, are distended and full. Symptoms of nasal stuffiness and facial fullness accompanied these physical signs, and persisted until the end of the mission. It has been estimated that 1.5 to 2.0 litres (2.6 to 3.5 pints) of fluid may take part in this cephalic migration, some of it remaining in the intravascular compartment (that is, as blood) but much of it entering and expanding the extravascular compartment as fluid moves out of the circulation.

But it is the increase in central blood volume which is believed to be responsible for the subsequent cardiovascular changes by acting upon, and resetting, all the controlling mechanisms. The end result is that a new circulatory status is achieved wherein total blood volume is reduced (by neural, hormonal, and renal mechanisms) and this, while being perfectly satisfactory for life in microgravity, leads to diminished tolerance to gravity on return to Earth. The manifestations of orthostatic intolerance range from a rise in heart rate and a fall in systolic pressure, with a consequent narrowing of pulse pressure (that is, the difference between systolic and diastolic levels), to an overt tendency to faint.

Both passive and active methods of assessing the degree of intolerance have been employed during the manned spaceflight programmes. The former has involved the measurement of cardiovascular variables such as heart rate, blood pressure, and limb volume, either when simply standing erect against a wall or when being tilted from the horizontal to a 70° head-up position; while the latter has employed the use of Lower Body Negative Pressure (LBNP). This technique involves the application of suction to the lower part of the body so that the resulting negative pressure causes displacement of fluid from the chest and head, thus simulating the effects of the erect posture in a 1G environment. The great advantage of the technique is that it can be used to assess the response to LBNP not only before and after flight but also, since it is essentially a closed vacuum system, during flight in microgravity.

The standing and tilt tests carried out on the crews of Gemini, Soyuz, and Apollo spacecraft revealed that, when compared with the normal pre-flight responses, there was a marked increase in response to these stresses post-flight, with exaggerated elevations in heart rate and leg volume, and a more profound decrease in systolic blood pressure. Similar results were obtained when these crew members were subjected to LBNP. The larger size of the Salyut and Skylab spacecraft allowed LBNP apparatus to be used on board to monitor the development of the cardiovascular deconditioning throughout a mission: for example, the response to LBNP at a suction of –50 mmHg (–6.7 kPa) was measured in all Skylab astronauts every three days. The first such tests were performed four to six days after launch and revealed an elevation in the levels of both resting and stressed variables, although the response was very unstable and remained so, especially during the first three weeks of a mission. As before, the post-flight response was also exaggerated, and it took several days to regain the pre-flight levels.

Several other indices of cardiovascular function have been assessed during spaceflight, including blood flow, venous compliance, and cardiac electric and dynamic behaviour. Blood flow through the legs was measured during the Skylab 4 mission and found to be significantly increased over pre-flight levels, probably as a consequence of reduced peripheral resistance and increased cardiac output. Similarly, venous compliance (that is, the ability of the veins to distend) in the lower limbs was markedly increased in flight although, having reached a peak ten days after launch, the magnitude of the change declined in later weeks of the Skylab mission. This increase in distensibility is believed to contribute to the exaggerated response to LBNP seen in space. On return to Earth, measurements of both blood flow and venous compliance returned to pre-flight values within hours. The heart size of all astronauts before and after flight has been compared by simple measurement of an X-ray image: and a small but consistent decrease noted, which may reflect some loss of cardiac muscle bulk (see also 'Haematology and Immunology', p. 166). Sophisticated techniques of assessing cardiac structure and function, including echocardiography (measurements of size and movement using ultrasound) and rheography (measurements of flow using devices to record changes in tissue resistance), have shown a decline in performance post-flight, with reduced efficiency of the left ventricle reflected as a reduced stroke volume. Echocardiography has also confirmed a loss in left ventricular muscle mass. Studies of the electrical behaviour of the heart, using standard electrocardiography (ECG) and the more specialized vectorcardiography, have demonstrated changes in the wave form and impulse conduction times of various components of the cardiac cycle. And, of course, routine monitoring of the ECG has revealed the presence of occasional dysrhythmias (see Chapters 2 and 9); while monitoring during the application of LBNP, and particularly just after its release, has likewise shown disturbances of conduction.

While these changes may be interesting, their full significance has yet to be determined. And fortunately, the evidence so far indicates that no major or irreversible decline in cardiovascular function occurs even after flights of many months' duration. Indeed, Russian evidence suggests that, while there may be small cyclical changes throughout a mission, most deconditioning has stabilized at its new set point after two to three months in space, and that readaptation on return to Earth is accomplished within weeks no matter how long the exposure to microgravity. The more immediate problem of post-flight orthostatic intolerance, and the possiblity of ameliorating its effects, is part of the wider issue of somehow preventing or reducing in-flight de-conditioning of all body systems, and is discussed on p. 180ff.

Fluid and electrolyte balance

On Earth, in healthy individuals, control of salt (sodium chloride) and water balance is carefully and accurately maintained by a series of neural, hormonal, and renal mechanisms; the intention being to maintain the constancy of both the volume and the osmolality of the extracellular fluid (see also Chapter 7). Although the controlling mechanisms for salt and water are separate, they are nevertheless interdependent. Thus, receptors (osmoreceptors) sensitive to changes in plasma osmotic pressure exist within the brain, while others (baroreceptors) within the cardiovascular system (and especially the left atrium) are sensitive to alterations in blood pressure, and yet more within the kidneys respond to changes in blood volume.

Should a water deficit exist alone, the osmoreceptors and baroreceptors stimulate the thirst centre at the same time as initiating the release of antidiuretic hormone (ADH) from the posterior pituitary gland at the base of the brain. ADH, which is also called arginine vasopressin, promotes retention of water by acting directly on the renal collecting ducts to increase their permeability, and so enhance reabsorption. In this way, smaller volumes of concentrated urine are excreted. Conversely, a water excess will reduce plasma osmolarity and increase atrial pressure, ADH release is inhibited and large quantities of dilute urine are passed – a diuresis.

A salt deficit, in the absence of a parallel decline in water, will also reduce plasma osmolarity. ADH is again inhibited and both plasma volume and total body water are reduced by a diuresis. The reduction in volume then stimulates receptors in the kidneys to secrete a hormone called renin which converts a circulating substance called angiotensin to an active form, capable of stimulating the release of yet another hormone, aldosterone, from the adrenal gland! Aldosterone acts on the renal tubules to retain salt and, because the salt carries water with it, fluid volume is restored as well, as ADH once more becomes active. Salt excess reverses the sequence of events: plasma osmolarity increases and water is retained (by ADH) in order to dilute the plasma and re-establish the status quo. The increase is blood volume inhibits the renin-angiotensin-aldosterone mechanism and the excess salt is excreted. The consequent fall in osmolarity promotes a diuresis and normal plasma volume is restored. A further hormone, the so-called natriuretic (sodium-losing) factor, has recently been identified which acts specifically to promote loss of sodium.

Microgravity profoundly affects the normal distribution and composition of body fluids, the changes being inextricably related to and largely directly consequent upon the changes within the cardiovascular system described earlier. It is an entirely reasonable supposition that the body would compensate for such alterations by invoking the usual mechanisms of Earth-bound salt and water homeostasis. Accordingly, the redistribution of fluid from the lower body to the chest and head should be interpreted by the stretch receptors in the left atrium as a real increase in circulating blood volume. This would then trigger the normal sequence of hormonal events to stimulate the kidneys to increase their excretion of both water and electrolytes. Evidence from in-flight studies does support this model in general terms, but no marked diuresis was observed during Project Skylab. Indeed, the daily urine volume *de*creased by an average of 400 ml during the first six days of the Skylab missions. But, since the average daily intake fell over the same period by 700 ml, a net loss of fluid did occur and this has been attributed to diminished thirst combined with losses in vomit as space motion sickness took its toll. (It will be recalled that dehydration has been a feature of most manned spaceflight programmes: see Chapter 7). Despite this slight anomaly, which was not helped by operational difficulties surrounding the accurate documentation of urine volumes during the early days of a mission, other biochemical evidence does support the predicted physiology. Thus, there was an in-flight increase in urinary loss of sodium, potassium, and chloride which was reversed on return to Earth. The associated and predicted fall in ADH was also observed, but the expected decline in aldosterone levels was not and remains an enigma. Aldosterone levels rose during flight and, while this would be consistent with potassium loss, it would not explain the fall in sodium. It may be that the natriuretic hormone is active under conditions of microgravity. On return to a 1G environment, the reductions in total body water and plasma volume are gradually recovered over a period of one to two weeks, and with them the normal electrolyte balance.

Haematology and immunology

Haematology is the study of the physiology of blood, including that of both its fluid (plasma) and solid (cellular) components. Plasma normally accounts for over half the blood volume and acts as the medium in which cellular components are conveyed about the body. Each litre of plasma also contains 65–80 gram of proteins which, besides having functions of their own, are responsible for the osmolality of the fluid. The cellular components of blood comprise red blood cells (erythrocytes), white blood cells (leucocytes) of various types, and platelets (thrombocytes). The principal function of erythrocytes is the carriage of oxygen to the tissues, while leucocytes are concerned with the defensive responses of the body to direct or indirect injury, and thrombocytes are essential for the blood clotting process.

The changes in plasma volume associated with microgravity have been discussed above, but changes are also seen in constituents of plasma and, particularly, in the structure and function of red and white blood cells.

The exchange of tissue nutrients and waste products occurs from and to the capillaries within each tissue bed. Capillaries are, therefore, relatively leaky vessels and, since the pressure within the interstitial compartment is approximately zero, it is clear that the pressure gradient normally present within capillaries would predispose to bulk loss of fluid (transudation) from the circulation unless prevented by some other mechanism. That plasma does not normally flood out of the circulation is self-evident, but the property which prevents such leaking is imparted to it by the presence of proteins which exert an inward component to plasma osmotic pressure, the so-called oncotic pressure. The plasma proteins (primarily albumin (55 per cent) and globulins (42 per cent)) are, therefore, vital to the maintenance of circulatory integrity. (The globulin fraction is also an essential component of the immune system – see below.) Should the level of such proteins in the blood fall, fluid would indeed enter the interstitial compartment with consequent swelling (oedema) of the affected part. This situation obtains in many disease states, but fortunately no major disturbance of plasma protein metabolism has been observed in those studied in space. Occasionally, however, a small post-flight rise in several globulin components, and a fall in another, has been noted.

The same cannot be said for the in-flight behaviour of cellular components, and changes in erythrocyte physiology have been reported since the very earliest space missions. In the healthy adult male, each litre of blood normally contains about 460 ml of erythrocytes, while the figure in females is slightly lower (410 ml). This value, when expressed as a percentage, is known as the haematocrit. Russian and American studies in microgravity have shown that, while the post-flight haematocrit remained normal or only slightly depressed when compared with pre-flight levels, there is a consistent fall in total red cell mass. This observation was first made following the Vostok and Gemini flights, but has also been seen in all programmes thereafter.

The phenomenon was extensively studied in the Skylab astonauts using a radioactive dilution technqiue. A small quantity of blood was removed and the red cells labelled with radioactive chromium before being re-injected. Further samples were taken thirty and thirty-one minutes later and their radioactivity assessed: the dilution of the labelled cells by unlabelled gave a measure of total red cell mass. These tests were carried out before and after flight and revealed a gradual decrease in mass of about 14 per cent over the first twenty-one to fifty days, followed by an equally gradual recovery whether or not the subject was still in space. Thus, for the short duration mission of Skylab 2 (twenty-eight days) recovery did not commence until two weeks after landing, while recovery had already begun by the time the crews of Skylab 3 and 4 returned to Earth. Russian physiologists have avoided the injection of radioisotopes for this purpose and instead measured haemoglobin mass spectroscopically: precisely the same pattern of change has been observed with a loss in mass of about 33 per cent after twenty-one to fifty days in space recovering to a 16 per cent loss after 100 days. Haemoglobin concentration within individual cells, however, has been normal or slightly elevated following short missions, but declined slowly throughout missions longer than fourteen days.

The reasons for the changes in red cell mass remain obscure, although it appears to be a suppression of red cell production in the bone marrow rather than the increased destruction of cells once in the circulation, since the life span of erythrocytes is unaltered in microgravity (as assessed by further isotopic studies). Certainly, the numbers of red cell precursors (reticulocytes) present in the blood follow the same pattern of in-flight decline and recovery as the red cell mass itself. The production of erythrocytes by the marrow is accomplished primarily by the influence of a hormone called erythropoietin (largely produced by the kidneys), the release of which is very powerfully stimulated by tissue hypoxia and inhibited by hyperoxia. At one stage, therefore, it was believed that the marrow suppression was a result of oxygen toxicity, but this is probably not so since it has been described in cosmonauts breathing sea level equivalent atmospheres, and persisted in astronauts despite the change in atmospheric composition for Skylab. It is now felt that the decline in red cell mass is a consequence of the fall in plasma volume, perhaps directly inhibiting erythropoietin, and is a response intended to maintain constant red cell and haemoglobin masses. Having reached its nadir after about one month, red cell mass does not then return to its pre-flight value for up to two months whether in space or on Earth, although the reticulocyte count recovers more quickly. The recovery, even while still in microgravity, is probably initiated by mechanisms related to the appearance of ageing red cells, since their normal life span is 120 days.

Changes are also seen in erythrocytes at the cellular level. Under the microscope, normal red cells are seen as biconcave discs (discocytes) and usually account for 80–90 per cent of the forms present. Of the remainder, some may be of various elliptical, bell-shaped, or spherical morphology, and some may have spines (echinocytes). Microgravity reduces the numbers of normal discocytes, and abnormal forms increase in number, particularly the echinocytes. Although potentially significant from a clinical aspect, these morphological changes are reversed within three days of return to Earth and there have been no apparent sequelae. The other important indices of red cell function, the mean corpuscular (cell) volume, the mean corpuscular haemoglobin, and the mean corpuscular haemoglobin concentration (see above), are variably but insignificantly affected. And other studies of erythrocyte metabolism, including the behaviour of cellular enzymes and lipids (fats), have all supported the conclusion that spaceflight does not induce irreversible red cell damage.

White blood cells (leucocytes) form the second group of cellular components in blood. Several different types of leucocytes are usually present in a recognized distribution and a normal differential white cell count shows granulocytes to be the most numerous (67 per cent) followed by lymphocytes (27 per cent) and then monocytes (6 per cent). Granulocytes (or polymorphonuclear leucocytes!) are formed in the bone marrow, have a life span of about thirty hours in the circulation, and are responsible for the body's non-specific cellular defence mechanisms. It is these cells which are mobilized to protect the body against bacteria, viruses, foreign particles, and even, on occasions, against some body tissues (for example, erythrocyte breakdown products). Their numbers increase dramatically during infections and other insults, but several different types exist (neutrophils, basophils, and eosinophils – depending on their histological staining characteristics) and each may dominate numerically according to the nature of the crisis. Thus, while they are all able to migrate to the site of an attack and are involved in its rebuttal, some cells are more appropriate than others in certain circumstances. For example, neutrophils and basophils are most numerous in response to a bacterial attack, while eosinophils are involved primarily with the response to allergens. The less numerous monocytes provide a second line of non-specific defence and, like their granulocyte cousins, are able to phagocytose (ingest) foreign material. The third type of white cell, the lymphocyte, is the essential element of the body's immune system.

Immunology is the study of resistance to infection and, in the human body, two major divisions comprise the immune system. The first, cellular component is mediated by one population of lymphocytes which, under the influence of the thymus gland early in life, acquire the ability to develop antibodies to foreign material, and subsequently to protect the body from a later exposure. These so-called T-cells (Thymus-dependent) are therefore responsible for the delayed immunological response to threats, including that from numerous external agents, transplanted material, and even some cancers. Although this last attribute is clearly beneficial, the mechanism may occasionally be perverted and the immune reaction directed against the body's own healthy tissue to produce an autoimmune state. Significantly, autoantibodies to cardiac muscle were demonstrated in the blood of 45 per cent of one series of cosmonauts and, although the response was lost after readaptation, this finding supports the cardiovascular evidence that heart size decreases in space and does suggest that cardiac muscle is not exempt from the general muscular atrophy induced by microgravity. Furthermore, the autoimmune response was particularly strong in those cosmonauts who had flown twice before.

The second component of the immune system is humoral (that is, fluid based rather than cellular) and is mediated by antibodies carried in the plasma. These blood-borne antibodies are proteins, and are specifically termed immunoglobulins. Immunoglobulins (Ig) are manufactured by the second, non-thymus-dependent, population of lymphocytes when they mature into plasma cells. Several different types of immunoglobulin exist, each with actions against specific threats: thus, for example, IgG has major anti-bacterial and anti-viral activity, and IgE is active in hay fever and against parasitic infections. Unlike the T-cells, the Ig system can initiate a response within minutes.

Since any impairment of the immune system is potentially fatal to the victim, the possibility that it could be adversely affected by spaceflight was obviously of great concern, and the crews manning both the Salyut and Skylab missions underwent considerable investigation in this field. A marked impairment of T-cell function was noted in crews returning from long duration missions, but normal behaviour was re-established within three to seven days and the clinical significance does not appear to be great. Similarly, only minor changes in some Ig elements were observed, again usually after intermediate and long-duration flights. Fortunately, therefore, both the cellular and the humoral immune systems in vivo appear to be largely unaffected by microgravity. There have, however, been some worrying results from certain in vitro experiments carried out during the Spacelab D-1 mission (STS 22) in October 1985. One study demonstrated that the immunological response of white blood cells, both in cultures prepared on Earth before flight and in whole blood extracted from the crew in-flight, was inhibited in microgravity. Two other studies revealed that bacterial cell proliferation was increased in microgravity *and* that their resistance to antibiotics was enhanced. If such findings pertain to living beings, the potential effects of increased susceptibility to infection by drug-resistant organisms could be devastating. And the threat of exposure to an unknown external agent while travelling in space will remain, of course, so that continued vigilance will be essential.

Finally, platelets (thrombocytes) form the third major particulate component of blood, and play a major role in haemostasis (clotting) where they are largely responsible for the physical nature of blood clots. There is no evidence from haematological studies to suggest that platelet function is degraded either during or after exposure to microgravity.

The musculo-skeletal system

The musculo-skeletal system comprises the bones and the muscles attached to them; the former give to the body its physical structure and strength, while the latter provide the means by which posture is controlled, movements are made, and physical tasks performed.

Mature bone is composed of inorganic materials (45 per cent), an organic matrix (30 per cent), and water (25 per cent). Most important among the inorganic substances are calcium and phosphorus, which make up the hard substance which is the characteristic of bone. A hard, dense external (cortical) layer gives the bone its shape, while internally a lattice-work of supporting trabecular bone is surrounded by the softer intercellular matrix. This gelatinous matrix is composed primarily of a protein called collagen, itself rich in a substance known as hydroxyproline.

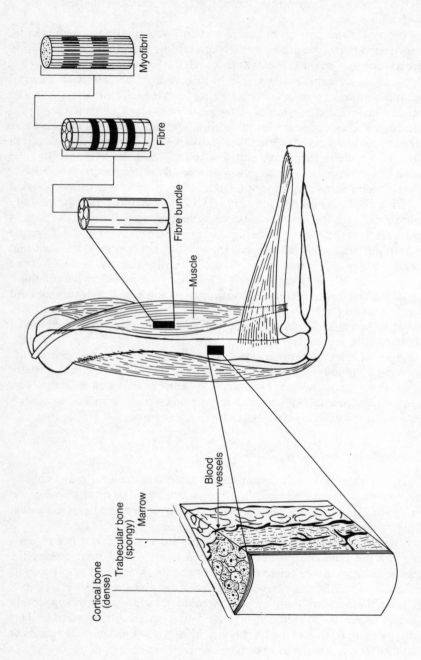

Figure 11.3 The structure of bone and skeletal muscle

A state of finely-balanced dynamic equilibrium exists in bone whereby there is a continuous process of resorption and deposition. The former is stimulated by various circulating hormones from the pituitary (soma-totropin), thyroid (thyroxine, tri-iodothyronine), and parathyroid (para-thyroid hormone) glands, and by vitamin D; and is inhibited by other hormones from the adrenal (cortisol and oestradiol) and thyroid (calcitonin) glands. The same hormones which inhibit resorption act to stimulate deposition as does phosphorous and, most relevantly, physical stress (including, therefore, gravity). Similarly, the substances which stimulate resorption will inhibit deposition. Thus, a balance of feedback mechanisms exists to maintain bone metabolism, and particularly that part played by calcium, at a level appropriate to the body's needs. Dietary intake, and urinary and faecal excretion, will also enter the equation, although for healthy individuals on a balanced diet these aspects are not critical (but see below).

Muscles are made up of many thousands of cells or fibres, each of which contains 1,000 to 2,000 contractile elements called myofibrils. A series of parallel rod-like structures within the myofibrils slide over one another under the influence of nervous stimulation, so shortening the myofibril and, with similar action occurring throughout the muscle, causing a contraction. Groups of muscles act together to produce graded and co-ordinated movement of joints. The strength of skeletal muscle is influenced by external physical factors, including the force of gravity against which muscle groups must work in order to achieve movement. Thus, the anti-gravity muscles of the limbs and spine may be expected to be developed to an extent which allows all normal physical activities to be performed. Athletes, of course, need to develop particular groups of muscles beyond this basic level in order to participate effectively in their chosen sports; and muscular fitness is correctly equated with general bodily fitness and health. When muscle tissue no longer has to maintain its own efficiency, however, as for example when illness confines an individual to bed, so-called disuse atrophy (wasting) occurs. The physical bulk of the affected muscles is reduced as their fibres shrink in size rather than diminish in number. Cardiac muscle, although similar in structure to its skeletal counterpart, is rather more specialized in that it has an intrinsic ability to contract (pacemaker function), and its response is all or none rather than graded. The smooth muscle of the viscera and blood vessels is different again, and can exhibit a graded response under neural influence as well as having a slow pacemaker function. Disuse atrophy is not believed to affect this type of muscle.

Clearly, therefore, the strength of bones must be such as to provide physical support to the soft tissues of the body when under the influence of Earth's gravity, and it is the force of gravity which dictates and maintains the mechanical strength of bone. Similarly, the strength of muscles must be such as to be able to move the body within the 1G environment. Once that force is removed, however, the necessity for mechanically strong bones and muscles, and the influence for their development, is removed and these tissues literally begin to melt away. The magnitude of muscle atrophy and bone demineralization is directly related to the length of time exposed to microgravity, and so has vital significance for any long-term mission into space; or rather, for the subsequent safe return to Earth of the voyagers.

The intermediate duration missions of Apollo and early Soyuz craft resulted in some loss of muscle bulk from the legs, and of some disturbance of gait just after return to 1G, but the changes were rapidly reversed and were not an important problem. Muscle atrophy was studied in more detail immediately before and after the Skylab missions, using a technique called four-camera stereophotogrammetry for biostereometric analysis! This rather grandiose title describes a method of taking sequential stereo photographs of the whole body, or of individual regions, and then analysing the pictures three-dimensionally to elicit any alteration in volume. Eight of the nine Skylab astronauts demonstrated a reduction in volume of the legs and trunk; a decline which was attributed partially to loss of muscle bulk and partially to the fluid shifts described earlier. There was a parallel loss of limb muscle strength, as measured after flight by another splendid-sounding instrument, the Cybex isokinetic dynamometer. This device, which was also used by the Russians, measured the peak muscle forces required to rotate its mechanism against a load at a pre-set speed.

The degree of muscle atrophy and loss of strength was reduced, but not eliminated, by using exercising devices (see 'Countermeasures' p. 180). And atrophy from the legs was always greater than that from the arms, presumably because the arms are involved in relatively greater activity in space. When muscle tissue breaks down, some of the metabolites produced are nitrogen-containing compounds and, since nitrogen can be readily measured in the urine, it is not surprising that the progressive muscle wasting was accompanied by an increased level of urinary nitrogen: the daily level being an average of 4 grams more during flight than before flight. (Normally, men excrete about 15 grams of nitrogen daily in their urine as urate, urea, ammonium, and creatinine.) Although the increased loss of nitrogen continues throughout flight, a return to pre-flight levels is achieved within days after landing.

On the basis of findings from ground-based studies of individuals immobilized in bed, it had been predicted that exposure to microgravity would lead to bone mineral (calcium and phosphorus) loss. This was confirmed by post-flight X-ray examination of the Gemini 4, 5 and 7 astronauts. Even though these flights lasted just four, eight, and fourteen days respectively, extensive demineralization of the radius (3–18 per cent), ulna (3–9 per cent), and heel (os calcis) (2–8 per cent) was demonstrated. (The original figures were even higher – up to 25 per cent loss from the radius – but some technical errors were detected later and the data re-evaluated.) Similar

X-ray findings were reported for the crews of Apollo 7 and 8, and the average post-flight loss from the os calcis for the twelve astronauts studied by X-ray densitometry was 3.2 per cent.

For later Apollo missions and for Skylab, a more sophisticated technique, photon absorptiometry, was adopted. This involved the measurement of gamma ray absorption by the bones under study, and again the distal (far) ends of the radius and ulna, and the os calcis were chosen. The very worrying findings from earlier studies were not repeated, and indeed no significant loss from any of the three sites was demonstrated in the crew of the twenty-eight-day Skylab 2 mission. Furthermore, no significant losses from the radii and ulnae were seen on the other two missions, although losses of 7.4 per cent, 4.5 per cent, and 7.9 per cent were recorded from the os calcis bones of the Skylab 3 (fifty-nine days) scientist pilot, and the Skylab 4 (eighty-four days) scientist pilot and pilot respectively.

Loss of mineral from weight-bearing bones has also been reported by Russian investigators, the extent being directly proportional to the mission length. The demineralization was considerable during the early flights of up to six months; and even the subsequent use of extensive countermeasures has failed to reduce the loss below about 3–8 per cent. In addition, demineralization is primarily from the internal supporting, or trabecular, bone rather than from the outer, cortical material. There is also a loss of material from the supporting collagenous matrix. Unfortunately, while trabeculae which have been partially eroded may regain their original dimensions (although post-flight recovery of bone mass takes about the same time as its in-flight loss), those which have been resorbed completely cannot be restored. Bones which have been weakened in this way are therefore liable to fracture.

Just as urinary nitrogen is an indicator of muscle breakdown, levels of urinary and faecal calcium reveal the status of bone metabolism. As explained earlier, calcium metabolism is usually in a state of dynamic equilibrium: a normal diet provides about 800 mg calcium daily (range 680–1,400 mg), of which 720 mg pass straight through in the faeces and 80 mg are excreted in the urine. A state of negative calcium balance develops in microgravity, however, beginning just after orbital insertion and continuing, unabated, until return to Earth when normal calcium metabolism is slowly re-established. Although diet was carefully controlled during the Skylab missions, urinary levels of calcium increased rapidly over the first month to reach a plateau for the rest of the flight at which the additional loss amounted to about 160 mg each day! Furthermore, calcium absorption by the gut was also affected and a gradual but sustained increase in faecal loss was observed throughout the mission. During the eighty-four-day mission of Skylab 4, the three crewmen lost an average of 25 grams (0.9 oz) of calcium from a total body content of about 1,250 grams (2.7 lb). If such a loss was sustained, about 25 per cent of body calcium would be lost during one year in space. It has been suggested that this rate of loss would result in clinically overt, and possibly irreversible, bone disease (similar to that seen when normal bone ages – a condition known as osteoporosis) following missions lasting more than five to seven months if no prophylactic (preventive) countermeasures are taken. Since a mission to Mars and back will take eighteen to thirty-six months, the implications of not doing so are obvious. Finally, the soft bone matrix also deteriorates under the influence of microgravity, and urinary hydroxyproline gradually increases throughout flight.

Body mass

A reduction in weight following space flight has been a constant feature of both the Russian and American manned spaceflight programmes. And, although several theories had been suggested to account for this, it was not until the prolonged scientific missions of the Salyut series and Project Skylab that the phenomenon could be investigated. Since weight is a function of the force of gravity, however, it clearly cannot be measured under conditions of 'weightlessness'! Instead, it is necessary to measure mass, which is unrelated to gravity, and is a property of inertia (that is, of the resistance of a body at rest to motion, or to a change in motion if it is moving). Although the terms weight and mass are often used interchangeably, the mass of a body, measured in kilograms, is constant no matter what gravitational forces of attraction from other bodies act upon it. Weight, on the other hand, is a numerical reflection of such forces and equals mass × gravitational force: weight can therefore vary, and on the Moon is one-sixth of that on Earth.

In microgravity, the mass of a body can be measured by assessing its inertial properties. The device used was a spring-mass oscillator, the principle of which was to record the precise timing and magnitude of natural oscillations induced by the body under test when set in linear motion between two springs. The mass of the body could then be derived mathematically from a knowledge of the characteristics of the springs, the displacement caused, and the period of oscillation. In use, the subject sat strapped into the device between the springs. He was also obliged to adopt a flexed and rigid attitude, and to hold his breath, in order to reduce errors from both the non-rigidity of his body and any external influences (known delightfully as slosh and jitter respectively)!

Measurements from the spring-mass oscillator confirmed the previous general finding of a 3–4 per cent weight (mass) loss during spaceflight. The magnitude of the loss, however, did not appear to be related to time spent in space, and in missions of less than six months' duration the average loss of 6–7 kg (13.2–15.4 lb) was independent of flight time. The pattern of loss was consistent, with a rapid fall over the first three to five days in microgravity, normally followed by a lesser but sustained decline over the rest of the mission. On return to Earth, there was an equally rapid restoration of much of the lost mass over the first few days, and pre-flight levels were usually regained within two to three weeks. The early phase of in-flight weight reduction has been attributed to the loss of fluids described above, as a result of both diminished intake and increased output (including the vomiting associated with motion sickness); while the later sustained decline probably reflected a combination of metabolic (dietary) imbalance, and muscle and bone loss. Some weight may also have been lost as a result of physical activity and heat stress. It was as an attempt to prevent the in-flight decline in body weight that the caloric value of the food provided for cosmonauts and astronauts was steadily increased as in-flight experience grew (see Chapter 7). But this manoeuvre can never be entirely successful since the other causes of weight loss will always operate in microgravity. Paradoxically, a gradual weight *gain* was observed in three of the cosmonauts who undertook six-month missions in Salyut 6. Evidence of fluid loss, and of muscle and bone breakdown, was still seen, however, and this aberrant finding has been explained by an increase in adipose (fat) tissue greater than the loss of lean body mass; that is, it was explicable in terms of dietary intake.

Exercise capacity

It is clear that physical fitness and the ability to exercise (that is, to undertake work) have played a vital role in man's success in space for, while microgravity could be expected to make some tasks easier to perform, the constraints imposed by spacesuits with reduced mobility, and the time needed to achieve a particular physical goal, meant that energy expenditure would inevitably be raised above resting levels for considerable periods (see chapter 7). Although the capacity of cosmonauts and astronauts to withstand these times of increased work could be improved by pre-flight training, it was predicted that the effects of spaceflight upon exercise physiology would degrade this capacity. Specifically, the ability of man to exercise is an important indication of his cardiovascular, musculo-skeletal, and pulmonary (lung) fitness. And microgravity has been shown unequivocally to degrade both the first and second of these, with a considerable recovery period required post-flight.

But a large number of pulmonary function tests performed on the Skylab astronauts immediately after landing, and subsequently, showed no significant change from pre-flight values. Vital capacity (see Chapter 4) was the only variable measured in-flight as well, and although a decrease of about 10 per cent was observed in all three Skylab 4 astronauts and one crew member on Skylab 3, even this had corrected within two hours of returning to Earth. The reduction in vital capacity has been attributed to a combination of the increase in body fluids present in the chest, a tendency of the diaphragm to be elevated in microgravity, and the reduced barometric pressure within the spacecraft. The net conclusion of all these discrete changes in cardio-

173

pulmonary and musculo-skeletal physiology was that exercise capacity would indeed be affected by spaceflight.

The body's response to the stress of exercise is to increase its metabolic rate to a level which matches the new energy demand (see also Chapter 7). Thus, both the cardiovascular and pulmonary systems are affected. Blood flow to working muscle is increased, at the expense of less immediately vital tissues such as abdominal organs and skin; although blood flow to the latter will rise if increased energy expenditure is prolonged to the point where excessive metabolic heat must be dissipated. The additional blood flow is achieved by a rise in cardiac output as a result of an increase in both heart rate and stroke volume. The additional demand for oxygen utilization and carbon dioxide elimination is met by increasing pulmonary ventilation, again by virtue of an increase in respiratory frequency and tidal volume (the amount of air moved in and out of the lungs during a single breath). Measurement of all of these variables, including metabolic gas exchange, will provide a quantitative assessment of the response to a given level of physical stress. It is, however, important that any method chosen for such assessment be repeatable and reliable so that serial studies can be made. Although many such methods exist, a graded stress protocol involving a cycle ergometer was chosen for the Apollo and Skylab projects since it could be used in-flight during the latter's missions. Thus, the subjects were tested on several occasions before and after flight, and during flight in the case of Skylab crews.

The cycle ergometer is a device with a saddle upon which the subject sits and pedals against a known resistance, and hence workload: the loads chosen represented light, medium, and heavy work as defined by heart rates of 120, 140, and 160 beats.min^{-1}. Additionally, loads at 180 beats.min^{-1} were imposed for the crews of Apollo 10 and 11. For each condition, the ergometer load was steadily increased until the pre-determined heart rate was achieved, thus providing a common end-point and a means of comparison. Workload, heart rate, blood pressure, and respiratory gas exchange (oxygen consumption and carbon dioxide production) were measured during each test of the Apollo astronauts, while cardiac output and stroke volume were also recorded during Skylab experiments.

The results from the Apollo studies, confirmed by Skylab experiments and those of the Russians, showed significant reduction in exercise capacity post-flight when compared with pre-flight levels. At a heart rate of 160 beats.min^{-1}, workload, systolic and diastolic blood pressures, cardiac output, stroke volume, and oxygen consumption were all depressed, although the mechanical efficiency of the body (the amount of oxygen required to perform a given amount of work) was unchanged. It took an average of three weeks for these changes to recover after return to Earth.

The in-flight results were somewhat different, however, and microgravity appeared not to affect the exercise capacity of Skylab and Salyut crew members. Furthermore, although the post-flight responses were depressed, the crews of the longer missions in each series have not, as may have been expected, required a correspondingly longer period to readapt. This may be a reflection of the increased in-flight exercise regimes adopted for these crews (see 'Countermeasures', p. 182).

The central and peripheral nervous systems

The brain and spinal cord together comprise the central nervous system (CNS), while the twelve pairs of cranial nerves, serving the head and neck, and the thirty-one pairs of spinal nerves, linking the rest of the body with the CNS, comprise the peripheral nervous system (PNS). The PNS has three major components: the motor nerves conveying commands to move *from* the brain and spinal cord to the skeletal muscles, the sensory nerves conveying information from the sense organs *to* the brain, and the parasympathetic and sympathetic complexes of the autonomic nervous system responsible for 'unconscious' visceral action such as respiration, digestion, and control of the cardiovascular system (see above). Through this vast network of nerves, the brain is able to control every conscious and unconscious activity of the human body. Because of its pivotal role, therefore, it is not surprising that neurological function while in microgravity has been of more than passing interest.

It is intuitively clear that man cannot survive and function in any normal way without a brain, and since it has been the purpose of this book to demonstrate that man can indeed survive and function in space, then *ipso facto* the brain must continue to behave normally in that environment. The higher functions such as perception, cognition, and memory, have all been undisturbed by spaceflight; except perhaps at the emotional level. And similarly, the basic body functions of which we are hardly ever aware have also not been unduly affected by spaceflight, given the protection of a spacecraft or spacesuit. Thus, respiration and circulation, mastication (chewing), deglutition (swallowing), digestion, metabolism and excretion (urination, defecation, and perspiration), and sensation, have all continued relatively normally with no major derangement or difficulty, either during or after flight. Only locomotion and some aspects of metabolism (that of bone and muscle), with the obvious effects of microgravity upon them, have been grossly affected. The changes, if any, in many of these aspects of human physiology have been considered earlier, but a number of relatively minor features of a neurological nature have been observed during exposure to microgravity.

Once in space, the absence of the force of gravity removes the constraints upon muscular action normally imposed on Earth. It might be expected, for example, that the decreased 'weight' of the hands and arms would affect the ability to perform precise movements requiring hand-eye co-ordination. Such an effect has indeed been demonstrated, in the form of past-pointing (the inability to place a finger accurately upon a selected point), immediately after entering microgravity. But the CNS very rapidly compensates for any overshoot or undershoot, and movements of fine control are accomplished without difficulty thereafter. Gross movements, such as those required for locomotion, are considerably aided by the virtual absence of a gravitational force; and, once techniques of movement are learnt, propulsion in this environment is achieved efficiently and easily. The large space stations of the Skylab and Salyut eras provided much photographic evidence of the normality of motor behaviour in space as routine daily tasks and experimental procedures, such as dressing (including the EVA suit), food preparation, general housekeeping, equipment maintenance, and use of the experimental hardware (cameras, LBNP devices, ergometers, etc.) were extensively filmed.

Some changes in neuro-muscular function have been demonstrated post-flight, however, and suggest that deterioration in performance at the cellular level has occurred in parallel with the general decline in muscle physiology described earlier. Thus, pre-flight and post-flight electromyographic (EMG) studies of the calf and thigh muscles have revealed a decline in muscular efficiency and a tendency to be fatigued: findings which mimic those of Earth-bound patients with diseases of muscular deterioration (atrophy). No such decline was demonstrated by EMG studies of arm muscles, however, so reinforcing the belief that the upper limbs retain their muscular health in microgravity by virtue of near normal levels of activity.

Further supporting evidence for the deterioration of muscle function comes from Skylab and ASTP studies of tendon reflex threshold (that is, the level of stimulation required to initiate a reflex action) and duration. A tendon reflex is a simple motor response, via the spinal cord (that is, it does not involve the brain and so is much faster), whereby stretching of a tendon causes the automatic (reflex) contraction of the muscle to which it is attached so preventing any over-stretch. The removal of, for example, the foot from a painful stimulus such as a nail, is a reflex phenomenon but is somewhat more complex since the response involves more than one muscle. These withdrawal reflexes are also protective: the painful stimulus causes automatic contraction of the limb flexors and inhibition of its extensors, both via the spinal cord, while the sensory impulse is also sent to the brain where it is consciously recognized. Other motor pathways activate the appropriate muscle groups in all the other limbs so preparing the individual for flight.

Pre-flight and post-flight studies of the Achilles (ankle) tendon reflex demonstrated sharp reductions in both threshold and duration of the reflex after flight compared with pre-flight levels, with recovery being related to mission length and taking over two months in the case of the Skylab 3 crew members. This increased excitability or exaggeration of a tendon reflex is termed hyperreflexia, and was seen in a generalized form in the crew of Skylab 2. As with the EMG changes, hyperreflexia on Earth is regarded as pathological and can be a feature of neurological disorders affecting motor neurones in the brain and spinal cord (that is, lesions of the upper motor neurones). The persistence of moderate to severe neuro-muscular dysfunction for several months after return to Earth, reported by the Russians after long Salyut missions, is further confirmation of a potentially worrying trend.

Incidental observations throughout the manned spaceflight programmes have indicated that the function of all the special senses remains normal in microgravity. Thus, the modalities of touch, taste, smell, and hearing are apparently unaffected by that environment. The same is true of the fifth and most important sense: vision. This is particularly significant since, as described earlier in this chapter, knowledge of body orientation, which on Earth is accomplished by a complex interaction of visual and vestibular mechanisms, is greatly disturbed by the degradation in the latter. In microgravity, therefore, the visual system is the most critical of all the senses for orientation, and hence for adaptation, to living and working in space. Because of this, visual performance was studied extensively throughout the early manned space flights.

Of particular concern was the ability of the eyes to cope with the increased luminosity (brightness) of objects viewed under solar illumination when in space, since light from the Sun is attenuated (absorbed) by at least 15 per cent before it reaches Earth by the normal atmosphere and the presence of clouds, water vapour, and any atmospheric pollution. The absence of light scattering also makes contrast even more pronounced, with the result that objects not being illuminated by the Sun appear darker than normal. The presence or absence of light in the surrounding environment affects the ability of the eyes to see objects within that environment clearly and accurately. This ability to resolve a visual target is known as visual acuity. Tests carried out before, during, and after the Gemini 5 flight demonstrated neither degradation, nor indeed improvement, in visual acuity in space. Similar findings have been reported by Russian scientists, and other tests of visual performance, such as contrast sensitivity and colour perception, have also been systematically investigated. Although the Russian studies suggest that there is a slight deterioration in visual function during the first few days in space, this subsequently recovers slowly and, in any case, appears not to diminish overall performance to any great extent.

Thus, both the American and Russian investigators conclude that visual performance remains as effective as it is on Earth. Even a 40 per cent loss in contrast sensitivity has not been regarded as a major problem in the environment of space. Some structural changes in the visual apparatus were demonstrated during the Apollo flights, however, although the significance of these, and of their slow recovery, has yet to be elucidated. Using retinal photography, vasoconstriction of the retinal blood vessels was observed in some astronauts, to a degree greater than that predicted by the presence of an oxygen-rich atmosphere. And some astronauts in the Mercury, Gemini, and

177

Apollo programmes were found to have reduced intra-ocular (eyeball) pressure post-flight.

Finally, some regard must be paid to the state of changed consciousness which occupies the average human being for over one third of his life: that is, to sleep. Sleep represents a readily reversible state of altered organization of the CNS during which sensory thresholds are elevated, muscular activity and reflexes are depressed, and visceral function reduced. Metabolic activity declines, as does core temperature, heart rate, and blood pressure, and respiratory behaviour is modified. In healthy adults, sleep is normally synchronized to the twenty-four hour cycle of night and day, and is entrained involuntarily. In volunteer subjects, however, in the absence of the external cues of dark and light, the body has a natural rhythm of about twenty-five hours to which the sleep-wake cycle adapts. Sleep itself is also a cyclic phenomenon which repeats a sequence of passing from light to deep stages and back again four or five times each night.

The different stages of sleep can be monitored by systematic recording of the electrical activity of the brain: the electroencephalogram (EEG). When the EEG is recorded with the eyes open, and the subject awake and alert, the electrical activity is dominated by beta waves which are of random occurrence (but at frequencies of fourteen to thirty cycles per second or Hertz (Hz)) and low voltage (5–10 millivolts (mV)). The EEG is then said to be de-synchronized. Removal of visual stimuli by closing the eyes allows the appearance of a synchronized EEG which is characterized by lower frequency, higher voltage alpha waves (8–13 Hz at about 50 mV). Drowsiness results in the reduced prominence of alpha waves and the appearance of slower waves. As the depth of sleep increases, the alpha rhythm disappears and is progressively replaced by beta type waves, slow theta waves (4–7 Hz), and so-called sleep spindles (short sequences of waves at 12–14 Hz). These features are themselves replaced in the deepest sleep by very slow delta waves of high amplitude (0.5–3 Hz at 20–200 mV). Every ninety minutes or so, however, this orthodox or dreamless sleep is interrupted by fifteen- to twenty-minute periods of light sleep during which rapid eye movements (REM) are evident on the electro-oculogram (EOG – measurement of electrical activity of the eye muscles). It is during this vital REM, or paradoxical, sleep that dreams occur, accompanied by some autonomic changes with the elevation of both heart rate and respiratory frequency, and an increase in cerebral blood flow.

Cosmonaut Titov in Vostok 2 was the first to demonstrate that it was possible to sleep in space, and evidence from his Russian and American peers supported the conclusion that such sleep was subjectively normal. Sleep during the early flights of Project Gemini, however, was not satisfactory in that periods longer than about four hours were impossible to achieve, and the crews of Gemini 4 and 5 (four and eight days respectively) were markedly fatigued post-flight. Several reasons for this were advanced, including the novelty and excitement of the experience, noise from within and without the spacecraft (and particularly voice communications from the ground!), altered circadian rhythms, staggered crew sleeping times, and general discomfort. As a result, for the fourteen-day mission of Gemini 7, further attempts to encourage normal sleeping patterns were adopted. These included structuring the mission plan on time at Cape Kennedy so that pre-flight biological rhythms were retained and sleep occurred during the hours of darkness at the Cape, and allowing both crew members to sleep at the same time so minimizing noise and general disturbance. Although these measures improved both the quantity and subjective quality of sleep on this mission, technical problems with recording equipment prevented objective analysis of the nature of sleep obtained.

During each Skylab mission, the EEG and EOG of one astronaut were successfully monitored during sleep. The results of these studies revealed that no major disturbances of sleep quality were induced by microgravity (Plate 16), although hypnotics were occasionally used to promote slumber. Difficulties have been encountered, however, with the readjustment of sleep cycles, and particularly the qualitative components, during the post-flight period. For many, the readaptation of sleep cycles to the 1G environment of Earth has proved to be more troublesome than adaptation to microgravity.

Although sleep in space has not presented any insoluble problems, its efficiency does depend upon the sensible use of such external influences as comfortable sleeping couches or beds, preferably in private areas separate from other activities, the maximum reduction of extraneous noise sources, the imposition of a familiar time-base on spacecraft activities, and the prudent use of medications if necessary. As with other aspects of advancing spacecraft technology and sophistication, crew sleeping quarters have been progressively improved. The early crews were constrained by cabin size to sleeping upon their couches. The Skylab astronauts and Salyut cosmonauts were provided with cot-like 'beds', in individual compartments, into which they strapped themselves in order to impart some feeling of security. And the Mir cosmonauts have been provided with personal cabins for private relaxation and sleep. The Shuttle mid-deck area contains a sleeping compartment for four: larger crews are therefore obliged to sleep in shifts, although extra sleeping bags can be attached to a locker bulkhead if necessary. Three beds in the compartment are horizontal but the lowest is used with its occupant facing floorwards, while the fourth bed is vertical: in microgravity, body orientation during sleep is clearly irrelevant! Each bed consists of a padded board 1.75 m (6 ft) long and 0.75 m (30 in) wide, to which is attached a sleeping bag. The bag has a full-length zip and an elasticated waistband which holds its user against the board with sufficient pressure to help create the illusion of sleeping on Earth. Privacy panels and curtains complete the dormitory arrangement. Eight hours are set aside each day for sleeping, with an additional forty-five minutes for preparation for bed, and the same period allowed after waking for washing and dressing. Ear plugs and sleeping masks are available, but if the entire crew is sleeping at the same time at least one astronaut must wear a communications headset in order to listen out for ground control messages and on-board alarms.

Countermeasures

The adverse physiological effects of microgravity, along with the progressively longer missions both undertaken and planned by the Russians and Americans, have led to a very great effort to establish effective and operationally convenient methods of palliation, particularly with regard to the ability of a returning crew to withstand the stresses of re-entry, landing, and the immediate post-flight period. Several of these countermeasures have been mentioned in this and earlier chapters, including physical intervention to help overcome cardiovascular deconditioning, the demineralization of bone, and the atrophy of muscles (Chapter 2 and above); dietary supplements to combat possible cardiac disturbances and renal loss of calcium (Chapter 9); and therapeutic manipulation to overcome the problems of space motion sickness and sleeplessness (above), and possibly even those of radiation (Chapter 5).

Since the cause of space deconditioning is the absence of gravity, it is not at all surprising that the most effective means of reducing its unwanted cardiovascular and musculo-skeletal effects has been the provision of extensive in-flight exercise regimes while under mechanical loading to simulate the force environment of Earth. This approach has been aggressively pursued throughout both the Russian and the American manned spaceflight programmes. Experience has shown that different forms of exercise, using different devices, provide protection against the various different components of deconditioning. Thus, loss of muscle mass and power is best minimized by isotonic exercises (in which muscles contract normally) and isometric exercises (in which muscles are prevented from contracting by acting against each other or against a fixed object); while simply standing or walking against an applied load reduces bone demineralization and muscle atrophy, and also improves post-flight co-ordination and gait; and endurance exercise such as pedalling an ergometer seems to maintain cardiac size and respiratory function but does little to help mineral and muscle loss. It is clear, therefore, that the optimum approach is to utilize a number of different techniques in an attempt to provide comprehensive protection. But there are some difficulties: for example, the time needed to be spent daily upon each form of exercise has not been established, and of course it may be that the ideal duration is totally unacceptable from an operational standpoint. Furthermore, the devices chosen must be easy and convenient to use and must be capable of operation within the space and weight constraints of the spacecraft involved.

A cycle ergometer has been a standard piece of equipment on board the large Salyut and Skylab spacecraft. It was described above (see 'Exercise

Capacity') and, as well as providing a useful experimental tool, allowed either the arms or legs to be exercised. Similarly, a treadmill has been provided for all Salyut space stations, for the Skylab 4 mission, and on board the Shuttle Orbiters. The user walks or runs on the device under conditions of simulated gravity induced by means of adjustable elastic cords (bungees). The cords are tethered at one end to the treadmill and attached at the other to a body harness worn around the waist and over the shoulders. The walking surface is made from a sheet of Teflon-coated aluminium.

Lower Body Negative Pressure (LBNP) has also been utilized by both the Russians and Americans not only as a means of assessing the development and degree of in-flight cardiovascular deconditioning, but also as a means of reducing it and so improving post-flight orthostatic tolerance. The principle and technique of LBNP was described earlier (see 'The cardiovascular system'), but the American hardware differed from that of the Russians. The device used on board Skylab was a static cylindrical chamber made of aluminium. It was about 1.22 m (4.0 ft) long with a diameter of 0.51 m (1.7 ft), and was designed to enclose the lower body of the astronaut, with a flexible seal around his waist. Salyut cosmonauts, however, retain a degree of mobility since they wear a vacuum suit termed the Chibis. This garment is really a pair of trousers, sealed around the waist, from which the air can be pumped. It can be used while sitting or standing, and has been employed particularly towards the end of long missions in an attempt to restore orthostatic tolerance. It is usually worn for fifty minutes each day for the two days immediately before landing, and for twenty minutes daily for the previous four days. To help counteract the fluid loss of prolonged spaceflight, cosmonauts also drink 300 ml (0.53 pint) of isotonic saline (that is, saline having the same osmotic pressure as body fluids) during the final LBNP session.

Another special garment, the 'Pengvin Suit' is also worn routinely by Russian crew members (see Chapter 2). This elasticated suit provides passive but constant loading against the anti-gravity muscles of the legs and trunk, and so provides partial compensation for the lack of gravity. A further type of suit, the anti-gravity (anti-g) suit, is used by cosmonauts and astronauts during the recovery sequence. This suit applies *increased* pressure (see the Chibis Suit) to the lower limbs and abdomen and so reduces peripheral pooling of blood and improves post-flight orthostatic tolerance. A number of accessory instruments and techniques have also been utilized at various times in various spacecraft environments. Many of these were simple spring-loaded devices which imposed resistance to movement, such as the Mark I exerciser used on Skylab 3 and 4, or stretch-spring arrangements, such as chest expanders (also called the Mark II exerciser for the Skylab missions). Although such tools are only useful for specific muscle groups, they are nevertheless simple, small, lightweight, and inexpensive; and so do have a place especially where size is a constraint. Finally, electrical stimulation of muscle groups, using a device called Tonus, has been employed by Russian physiologists in an attempt to minimize atrophy. The effectiveness of this technique has not been established.

Table 11.2 summarizes the items of equipment used and exercise regimes adopted in the various long-duration manned spaceflight programmes.

Table 11.2 Summary of exercise devices used during certain manned spaceflight programmes

Spacecraft	Device	Combined Daily Use (hours)
Soyuz/Salyut	Exerciser Ergometer Treadmill Penguin suit Chibis (LBNP) Anti-g suit	2.0 – 2.5
Skylab 2	Ergometer LBNP chamber Anti-g suit	0.5
Skylab 3	+ Mk I exerciser Mk II exerciser	1.0
Skylab 4	+ Treadmill	1.5
Shuttle	Treadmill Anti-g suit	0.25 – 0.5

As experience in in-flight mechanical countermeasures has increased and their use refined, the undoubted benefits of exercising in space have become evident. There is no doubt that a sustained programme of physical exercise has improved the post-flight performance of the cardiovascular and muscular systems, and has maintained overall physical fitness. Thus, the results from the three Skylab missions revealed that, although the flights were of progressively longer duration, the increased exercise regimes adopted for the longest mission were able to reduce post-flight recovery times to less than those observed for the shorter missions. In addition, the previously observed post-flight deteriorations in cardiac output, stroke volume, orthostatic tolerance, leg volume, and leg strength were markedly lessened by the exercise manoeuvres adopted. The loss of bone mineral, however, was not greatly reduced. The Russian findings from the Salyut missions have been similar, with little change in post-flight exercise capacity, body mass, and response to LBNP when compared with pre-flight values.

The sustained loss of body constituents, despite the mitigating effects of exercise, clearly requires the provision of a calorifically adequate and nutritionally appropriate diet; and this too has been actively sought. Details of these aspects of food management systems were described in Chapter 7. In addition, dietary supplements in the form of extra vitamins, amino acids, and trace elements have been used routinely by cosmonauts and on occasion by astronauts. And potassium supplements were provided for the crews of Apollo after the in-flight cardiac dysrhythmias of Apollo 15 were attributed to potassium lack. There is, however, no evidence to suggest that this was of any benefit.

With the exception of those used for motion sickness, drugs used routinely as countermeasures (as opposed to their accepted use for medical indications) have not been very popular or successful because much basic research is still required before knowledge of, and confidence in, their actions is adequate. Russian physicians have favoured this approach rather more than their American colleagues, however, and have administered many different drugs (including hormones, steroids, and potent cardio-active and vaso-active preparations) at various times in an attempt to influence fluid balance mechanisms and the behaviour of the cardiovascular system. None of these drugs has been conspicuously beneficial. Efforts to minimize the loss of calcium from bone by pharmacological means have also met with little success; and certainly additional dietary calcium and phosphorus appear to have as little influence in space as they do in bed-rest experiments. The results from recent American studies of a diphosphonate drug (clodronate disodium), which reduced calcium loss by inhibiting resorption from the bones of those confined to bed, were encouraging until it was noted that the drug caused kidney damage. The use of anabolic steroids is also being investigated but, again, side effects are a problem. Because of the great resistance of bone metabolism to both external and internal efforts to reduce the effects of microgravity upon it, it is likely that the only completely acceptable solution will be the provision of artificial gravity while in space. This may be accomplished either by rotating part of a space station continuously or, if this is operationally undesirable, by the intermittent use of on-board centrifuges. Such devices have already been successful in preventing decalcification in rats after nineteen days on board a Russian biosatellite (Cosmos 936) in 1977.

Summary

Notwithstanding the considerable limitations of human experimentation in space so far (the small sample size, limited opportunity for study, and extensive use of countermeasures), it is quite apparent that there are considerable physiological changes associated with existence in microgravity, and equally important changes evident on return to Earth. Figures 11.4 and 11.5 are graphical representations, in a form developed by NASA but based upon both American and Russian findings, of the changes identified in various body systems during adaptation to space and then during readaptation to 1G respectively.

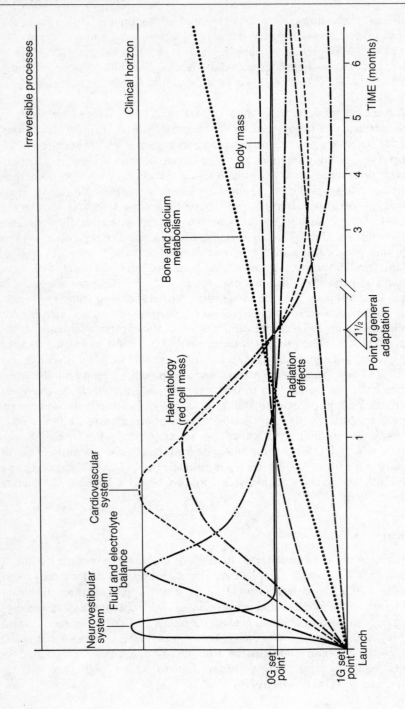

Figure 11.4 Physiological adaptation to microgravity (for explanation see text)

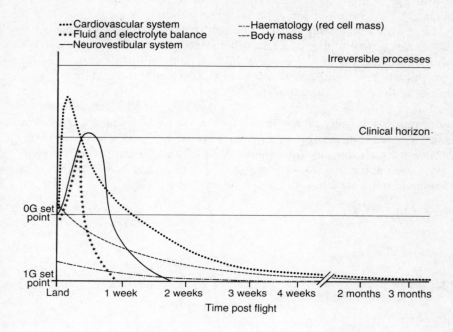

Figure 11.5   Physiological readaptation to the Earth's gravitational field (for systems largely independent of mission duration)

In both figures the curves represent changes in system processes and *not* the direction of such changes. Furthermore, all four horizontal axes are significant. The 0G and 1G set point lines with time are, respectively, the levels at which body systems are adapted to life in space or on Earth. The clinical horizon represents the line beyond which symptomatic changes will become apparent and the victim may be regarded as overtly 'ill'. In microgravity, for example, the clinical horizon is cut by the neuro-vestibular curve and this represents the appearance of space motion sickness. Similarly, on return to Earth, the cardiovascular curve cuts the line and indicates the occurrence of orthostatic intolerance. The irreversible processes line is self-explanatory, and the inexorable progress towards it of the curves describing bone and calcium metabolism, and radiation effects, emphasizes the importance of the influence of space upon these systems as described above and in Chapter 5. Of course, there is very considerable variability between the physiological responses of individuals and their speed of adaptation to a new force environment, and so these curves are idealized concepts which can only represent trends. But they are useful for illustrating the present state of knowledge of physiology as it relates to spaceflight, and they also serve to highlight the areas which are of particular concern.

So, most body systems are able to adapt to microgravity over a period of about one to two months, while post-flight readaptation is somewhat quicker with most systems attaining their 1G set point within four weeks. An exception to this was the persistent neuromuscular postural dysfunction seen in one of the cosmonauts from the 237-day mission of Soyuz T10. No other sustained, progressive, or residual pathology has so far been demonstrated in returning cosmonauts or astronauts, but the in-flight changes observed in some systems indicate that complacency would be dangerously misplaced. Active research into both the mechanisms of and the countermeasures to space deconditioning is still required if prolonged spaceflight *and* a physiologically safe return to Earth is to become a reality.

Chapter twelve

The psychology of spaceflight, women in space, and spacecraft habitability

The psychology of spaceflight

Our adventure into space has so far been dominated by the need to ensure physical survival in that hostile environment. Understandably, this has led to intense observation and investigation of physiological responses, and to the maintenance of a survivable milieu: the fruits of this successful effort are described throughout the earlier chapters of this book. Parallel studies of the psycho-physiological response of the human mind while in space have not been extensive (mainly because of operational constraints, but also because aircrew have an innate mistrust of psychologists and psychiatrists!), although such correlation would undoubtedly have been of great value and would have added further meaning to the mass of 'pure' physiological data gathered. In addition, while the physiological ability to survive in space has been established, the ability to adapt to the behavioural stresses of space travel has not, and this may yet be responsible for the success or failure of future prolonged missions.

Psychology is the science of the normal nature and function of the human mind; while psychiatry is concerned with mental *illness*. The distinction between the two disciplines is necessarily blurred but, clearly, a normal psyche is an essential feature of a healthy existence just as illness of the mind is to be avoided if at all possible. This section is concerned with those aspects of manned spaceflight which may be expected to affect normal psychological processes in a manner which may precipitate the appearance of abnormal mental states. Such problems include those of isolation and confinement, sensory overload, altered biological rhythms, and group interactions. Thus, part of the initial psychological examination of candidates for space travel was specifically designed to determine the extent, if any, of individual propensities to these types of stress (see Chapter 10). And this, combined with continuous reassessment during training, was the first stage in an essentially preventive approach to in-flight problems since, once in the unique environment of space, other potentially damaging psychological insults may occur without hope of immediate resolution.

Isolation and confinement

On Earth, when an individual is isolated from his normal environment, whether alone or in a small group, a set pattern of responses is seen, the intensity of which increases with time. The most common symptom is a disturbance of sleep, but other somatic complaints (such as headache, irritability, and anorexia or loss of appetite), anxiety, depression, boredom, restlessness, anger, and temporal and spatial disorientation are also reported; and, perhaps most significantly, there is a decline in task performance in parallel with this overall reaction. When others are present, several other factors may influence the overt manifestation of these features, including the desire to maintain at least a semblance of goodwill towards and co-operation with the rest of the group, the continued integrity of which may govern its very survival. This suppression of normal tension may of itself create problems. Such findings have been reported consistently from studies of groups of men isolated in submarines, on Polar expeditions, and in closed simulators within laboratories. It was therefore felt that isolation, confinement, and sensory deprivation would present a major psychological problem for men in space. Fortunately, this has not been the case so far, although many of the features described above were evident in Cosmonaut Vasyutin before his premature repatriation from Salyut 7. With the possiblity of very long space missions involving many crew members of disparate personality types, however, the potential for serious discord among crew members as a consequence of the environment of space must be considered (see 'Group Interactions', p. 190).

Sensory overload

With hindsight, it is perhaps not surprising that sensory over-stimulation of cosmonauts and astronauts has been of more significance during short and medium duration space missions than the expected problems of sensory deprivation. The uniqueness and excitement of the experience, the multitude of tasks performed, and the disruption of normal work/rest cycles have all conspired to produce a state of hyperarousal. Individuals so stimulated display a range of symptoms and signs which include sleeplessness and anorexia, and it will be recalled that these complaints were common, especially during the early manned spaceflight programmes. Furthermore, should the state of hyperarousal be prolonged, other manifestations may develop, including depersonalization, perceptual difficulties, mental and muscular incoordination, and an overall deterioration in performance; and indeed, it has been suggested that the unexpected and bizarre behaviour patterns exhibited by several cosmonauts and astronauts, which threatened the success of some missions, were attributable to these features of sensory overload. Predictably, haemodynamic changes, such as hypertension (high blood pressure), a rise in cardiac output, and a redistribution of blood flow from skin to muscle, are also a feature of hyperarousal and it has been postulated that, once the aroused state subsides (as it would, for example, after the undoubted mental stimulation of the re-entry and landing sequence), the reversal of these changes contributes to the immediate post-flight cardiovascular instability described in the previous chapter. Thus, the manifestations of hyperarousal may form a component of at least some of the overall physiological responses to spaceflight.

Biological rhythms

Virtually all living beings display regular variations in both behaviour and function with time of day, light and dark, temperature, season, and other environmental influences. Although these variations are governed or entrained by external factors (so-called zeitgebers – 'time givers'), the internal biological clock is responsible for the controlling mechanisms, which are probably hormonal in nature. In the human, over fifty such physiological and psychological rhythms have been identified which are linked to the twenty-four-hour day-night cycle and are therefore termed circadian rhythms. Most prominent among these endogenous (internal) cycles are sleep-wake periodicity (see Chapter 11), eating, and changes in autonomic functions such as blood pressure, pulse rate, respiration, urine volume, and body temperature. In the absence of any zeitgeber, the human internal clock extends its rhythmicity to a natural cycle of about twenty-five hours; while disruption of normal zeitgebers, as for example when crossing several time zones during conventional flight, leads to a corresponding desynchronization of the clock and to the symptoms and signs of the syndrome known as 'jet lag'. As with the sensory insults described above, this syndrome can give rise to sleep disturbances, alteration of eating and bowel habits, general discomfort and malaise, and a reduction in psychomotor performance (that is, performance characterized by body movements associated with mental activity). As a general rule, re-entrainment of circadian rhythms takes about one day for each one-hour time zone crossed; although westward travel, with its 'prolonged day', tends to be better tolerated than eastward as a consequence of the natural tendency of the internal clock to run at a longer daily cycle.

For the space traveller, the powerful and terrestrially omnipresent zeitgeber of day and night is obviously absent because the rotation of the Earth is no longer an influence. Furthermore, time zones become meaningless since each may be traversed in minutes by an orbiting spacecraft. The influence of clock hour does remain, however, and the work/rest cycles of crew members have been maintained by their internal clocks as entrained by the time patterns established before launch. Indeed, such synchronization has been imposed upon both Russian and American crews from the early days of the manned spaceflight programmes (see also Chapter 11). The adherence to Earth-based cycles has proved effective in preventing symptoms of circadian disruption not only for long-duration orbiting missions but also for the medium-duration lunar missions of Apollo. But it remains to be seen whether such methods will be either successful or desirable when *very* long-duration flights are undertaken, and when a permanent existence on planets with fundamentally different environmental characteristics to those on Earth is contemplated. As Dr Hubertus Strughold, regarded as a founding father of space medicine, wrote: 'The physiological clock, manifested in terrestrial circadian rhythmicity, will play an important role in health and performance in man's conquest of space.'

Group interactions

It is an unfortunate feature of life that individuals, even of the same species, are incapable of living and working together in peaceful harmony for any continuous period. Throughout history, this trait in humans has been manifest on a grand scale as a progression from one conflict to another between tribes, nations, and alliances of nations; and on a smaller but daily scale in domestic life. The behaviour of small groups of people, especially when in an isolated and/or dangerous environment will inevitably lead to increased tension which may or may not erupt into frank hostility. Space travel is an example par excellence of such a scenario. Group dynamics, or the ways in which individual members of the group relate socially to each other, are difficult to assess since they involve a complex interaction between the personality, status, role, and experience of each person.

The early astronauts and cosmonauts formed a remarkably homogeneous group with like physical and mental attributes (see Chapter 10) and, consequently, relatively little attention was paid to the inter-personal relationships and group interactions of the crews manning the pioneering space missions of the 1960s and early 1970s. As mission duration increased, however, more emphasis was placed in these areas upon both pre-flight assessment and monitoring of in-flight behaviour; especially by the Russians. Indeed, it has been a feature of the Russian manned spaceflight programme that considerable attention has always been paid to the behavioural aspects of space travel. Cosmonaut training includes extensive psychological testing, and crews are rostered according to compatibility. The advent of large, continuously manned complexes such as Mir and Space Station, crewed by many people with different backgrounds and motivations, means that behavioural psychology in general and group dynamics in particular will inevitably be of great importance. It may well become necessary to include formal training sessions in group behaviour prior to flight in order that psychological conflicts can be dealt with appropriately and without fear of jeopardizing either the safety of personnel or their mission.

Not least among the behavioural (and physiological) aspects of space flights are the potential repercussions of including both men and women crew members: a clearly desirable and necessary strategy in the world of equal opportunities, and an obvious requirement when colonization of other planets is contemplated.

Women in space

By early 1986, two Russian and eight American women had joined the elite corps of space travellers. Valentina Tereshkova was the first to do so twenty-three years previously in Vostok 6, but it appears that the Russian use of women in their manned spaceflight programmes has been as little more than publicity coups: there is no formal female cosmonaut cadre. Indicatively, Svetlana Savitskaya was met with a gift of an apron from her male colleagues on arrival at Salyut 7! American women, however, are now an integral part of the NASA mission-specialist astronaut training scheme; and women from a wide variety of scientific disciplines, including medicine, physics, engineering, geology, and biochemistry, have taken their place alongside similarly qualified male colleagues. And, of course, the first private American citizen to be selected for space travel was a female teacher. Female members of spacecraft crews are now a permanent and welcome feature, although male chauvinism (or chivalry ?) will certainly continue to exert an influence when pioneering and dangerous enterprises are undertaken: inevitably, for example, it has been announced that the crew of the first Shuttle mission after the Challenger disaster will be male. Apart from the historical precedent which dictates that men are the most experienced pilots, however, there is no valid physiological, psychological, or intellectual reason why this should be.

Notwithstanding the obvious physical differences between the sexes, most female physiology is qualitatively very similar to that of the male. It is, therefore, entirely predictable that, with regard to most stresses of spaceflight, the fairer sex will not be in any way disadvantaged. Experience so far has confirmed this belief. Thus, for example, men are no less susceptible than women to the problems of hypoxia, barotrauma, temperature extremes, exposure to radiation, nutrition and energy expenditure, and microgravity. There are some minor differences, however, but none which would debar an otherwise healthy woman from space flight. When compared with males, there is a small increase in the susceptibility of females to decompression sickness, probably as a consequence of the greater amount of adipose (fat) tissue present in their soft tissues (see Chapter 3); and centrifuge studies of female pilots have shown that increased +Gz accelerations induce both urinary stress incontinence and menstrual flow. The significance of these findings for the relatively benign re-entry acceleration profile of the Shuttle has not been established, and female tolerance to +Gz acceleration is otherwise comparable to that of males. Indeed, for some reason, female tolerance to such acceleration is significantly higher during menstruation.

Not surprisingly, the response to microgravity is also similar in both males and females, and the changes described in chapter 11 appear to affect both sexes in a like manner. It must be remembered, however, that the opportunity for study has been limited since very few women have been into space and certainly not for any great length of time: the longest so far, Savitskaya's second flight to Salyut 7 lasted just twelve days. Susceptibility to motion sickness, and to the entire Space Adaptation Syndrome, is similar in both sexes, but it possible that prolonged flight will reveal an increase in the magnitude of musculo-skeletal deconditioning in females. This is because, while physical fitness in both men and women has been shown to improve tolerance to environmental extremes such as acceleration, heat, hypoxia, and hypobaria (lowered atmospheric pressure), women do not develop the same degree of muscular strength or exercise capacity as men, even when trained in comparable physical fitness programmes. Furthermore, women on Earth are generally more prone to osteoporosis and so many demonstrate increased mineral loss from bones when in microgravity.

The unique area of female physiology is that concerned with gynaecological function. Prolonged occupation of space by humans will impose the need to reproduce outside the environment of Earth. Before this can happen, however, men and women will have to live together in microgravity under conditions conducive to sexual activity, conception implantation, gestation, and delivery! With regard to the first of these, one (female) psychiatrist particularly concerned with women in space has stated, 'Sexual intercourse in the weightless environment will undoubtedly raise very interesting possibilities.' Of course, much study into the effects of microgravity upon gynaecological function will be required before this stage is reached, and it will be essential to establish that the normal cyclical hormonal changes associated with ovulation and menstruation, and the processes themselves, proceed as on Earth. It will also be necessary to investigate methods of contraception in space, and later the vast area of the effects of conception, normal pregnancy, and birth on both the mother and child. And, in years to come, child development in space may become a reality. But in the immediate future the subsequent reproductive ability of men and women who have been in space may provide a rich area for research. The healthy daughter of cosmonauts Tereshkova and Nikolayev is good early evidence that post-flight normality is likely, at least as far as short duration space flights are concerned.

It is in the field of psycho-social behaviour, however, that the introduction of female crew members is likely to have the most immediate effect. In space, naturally enough, men will retain their masculinity and women their femininity, and it is inevitable that the same cultural stereotypes will obtain in that environment as on Earth. Thus, there will be a psychological impact upon the social structure, function, and behaviour of a mixed crew by virtue of it simply being mixed. Although on very long missions all three aspects may evolve in a manner different to that normally expected on Earth, sexual tensions and problems of leadership, especially when a female has a commander's post, will be added to the psychological stresses already discussed. The employment of married couples as crew members may be one method of defusing tension.

Finally, the presence of women on board has led to a reappraisal of several aspects of spacecraft design. In general, women are lighter and shorter than men, with less muscle mass and strength, and a shorter reach. Equipment must, therefore, be capable of operation by the full anthropometric range of both men and women: in most comparative body dimensions, the female 95th percentile is approximately equal to or less than the male mean (50th percentile). The sizing of garments, and particularly of spacesuits, must be appropriate to the needs of both sexes and this has led to the very large number of off-the-shelf components required for the present Shuttle spacesuit (see Chapter 8). Arrangements for toilet facilities must also bow to the needs of female anatomy. The design of Salyut, Mir, and Shuttle lavatories is such as to accommodate both sexes in private, while a disposable waste containment system, similar to the Apollo FCS, has been designed for use by women when wearing spacesuits (see Chapter 7). The flow and collection of menstrual loss, which on Earth is obviously influenced by gravity, is a further potential problem which will require consideration.

Spacecraft habitability

Habitability is concerned with the nature and quality of the environment and with the ease with which its occupants can adjust to it and operate efficiently within it while maintaining a sense of well-being and high morale. Accordingly, implicit in the study of human function in space must be the impact of certain external facets of normal daily existence. Many such aspects can be identified, including several which have been described and discussed in earlier chapters:

- Internal environment (atmospheric composition and revitalization, temperature, humidity, radiation, and ambient noise and lighting levels)
- Food and water (selection, storage, preparation, and consumption)
- Personal care, health and hygiene (waste collection and disposal, medical facilities, and grooming)
- General housekeeping (disposal of rubbish, cleaning and laundering)
- Clothing
- Mobility, ease of use of equipment, and restraints.

But several more, considered below, will be increasingly important as mission durations are extended:

- Spacecraft architecture
- Off-duty activities
- Personal privacy
- Communications.

All of these external factors can affect both physiology and psychology, and may be regarded collectively as the problems of spacecraft habitability.

The study of the efficiency of people within their working environment is called ergonomics (although it has come to be equated with the *design* of the environment (the man-machine interface) such as to optimize the performance of the people within it). The history of aviation is punctuated by episodes which demonstrated the need for great care in the design (shape, size, colour) and placement of instruments, controlling devices, and indicator dials. In very early spacecraft, the crew was held virtually immobile within the cabin, and ergonomic considerations were confined to ensuring that flight controls could be reached and operated adequately, and that indicators of vital spacecraft function were clearly seen and unambiguous. As the size and complexity of spacecraft increased, so the need for greater thought in the design of the workspace became evident; all the while influenced by the lack of true 'up' and 'down' in microgravity. Consideration must now also be given to ease of access and escape, to the location and dimensions of storage facilities, and to the overall geography of the craft so that movement of people and hardware is safe, economic, and unhindered. As an example of poor design, air flow within the early Salyut space stations passed from the toilet and wash areas (dirty) to the food and water storage areas (clean): not surprisingly, increased counts of gut bacteria were cultured from oral and nasal swabs taken from crew members!

Spacecraft architecture extends beyond the ergonomically efficient, however, to include aspects of aesthetic importance. Thus, interior decoration and arrangement become important when sojourns of many months are planned. Relaxation areas are as relevant as the primary working areas, and it is significant that the Russian designers have located individual cabins, with a bed, an armchair and a desk, for each crew member aboard the Mir space station. The Skylab, Salyut, and Shuttle crewmen also enjoyed some degree of privacy while sleeping, but it is only possible to provide more extensive facililities on the large space stations.

Despite the gregariousness of man, some privacy of both time and space is a cherished commodity. A place to which to retreat and in which to undertake private contemplation, writing, and reading, as well as personal grooming, is a council of perfection but one which must be heeded for prolonged missions. Private cabins are also the place for personal memorabilia, such as family photographs and other momentos of Earth, which will assume immense importance for the space traveller in the months and years away from home.

More public relaxation can be achieved within an area set aside as a form of wardroom, as on board the larger spacecraft of the 1970s and 1980s. Here the crew can gather while off-duty to prepare, serve and consume food (itself an important social ritual), to play card-games and board-games, to watch video tapes for entertainment, education and communication (such tapes are sent regularly to cosmonauts from their families back on Earth), and simply to converse. Personal music centres and pre-selected books complete the more obvious attributes of the off-duty environment. And it may well be possible to undertake even more ambitious hobbies, such as painting and gardening, on board space stations of the future. Facilities for exercise must also be provided, not only for the physiological reasons described in the previous chapter but also for the proven psychological benefits which accrue from being physically fit and maintaining that state. Predictably, Earth watching has proved to be an extremely popular pastime for those in orbit, and adequate observation windows must form an essential part of any spacecraft design.

Finally, of course, it is absolutely vital that communication between crew members on board spacecraft is maintained at all times, and especially during EVA. It is equally vital to ensure that the ability to communicate with Earth is retained for as long as possible. In this respect, it is just as important for crew members to be offered regular opportunities to speak privately to family and friends as it is for them to speak to the professional members of the ground control team.

Chapter thirteen

The future

There can be no doubt whatsoever that we shall continue to explore space, for our insatiable curiosity and quest for knowledge will drive us on as it has for centuries. And it must be almost inconceivable that, in the entire Universe, only the Earth has been populated by intelligent life forms: contact with occupants of other worlds may produce untold mutual benefits, and such exciting possibilities surely act as a spur. Even if this is not so, there may come a time when conventional life on the surface of the Earth is no longer possible for the entire population: the ocean floor may provide one haven, but travel to and colonization of other habitable planets may become an inevitable quest. As the Russian space pioneer Konstantin Tsiolkovksy once wrote: 'The Earth is the cradle of mankind, but one cannot remain in the cradle forever.'

Indeed, one recent survey of space scientists has predicted that large groups of people will be living in space by the year 2024, the first manned mission to Jupiter will take place by 2029 with rapid transport within the solar system by 2040 and a manned journey to its edge by 2058. The first manned interstellar probe may have departed by 2140, and colonization of an extra-solar planet is possible by the year 2260! Before these fantastic achievements are possible, however, rather more sedate progress must be made closer to Earth, and this chapter is concerned with those aspects of survival which will require especial attention or re-examination in order that man can undertake the prolonged missions within our own solar system proposed for the near future.

The subjects discussed in previous chapters will clearly be of as much concern to the survival and well-being of future space travellers as they were to those who pioneered manned spaceflight. In some cases, such as the need for improved protection against radiation (Chapter 5) and the requirements for greater consideration towards behavioural aspects (Chapters 10 and 12), the possible direction of future research and development has already been suggested. In others, such as the progressive and desirable reduction in launch and re-entry acceleration profiles (Chapter 4) and the design of clothing for life in space (Chapter 8), it is difficult to visualize any major or novel improvement or alteration in the proven technology (with the notable exception of spacesuits). In yet others, however, a brief consideration of necessary refinements is required.

Of most immediate interest in this respect is the manner in which the lessons of the past twenty-five years and present predictions are translated into the design of space stations in Earth orbit. Although the technology and skill already undoubtedly exist to support continuously manned orbiting craft such as Mir and Space Station, it is essential that fundamental improvements are made in the science of life support systems (Chapters 3, 6, and 7), in the approach to the resolution of the physiological problems of microgravity (Chapter 11), and in the provision of sophisticated in-flight health care (Chapter 9). This is because, in decades to come such stations will be regarded as mere stepping stones on the path to other worlds. They will serve as the launch sites for deep space missions in craft whose occupants will require vastly improved facilities to ensure survival both in space and on return to Earth.

Future life support systems

During each day in space, the average human will consume about 2.5 kg (5.5 lb) of water, 0.9 kg (2.0 lb) of oxygen, and 0.675 kg (1.5 lb) of dried food. He will also produce about 2.5 kg of urine, 0.8 kg (1.8 lb) of carbon dioxide, and about 0.2 kg (0.44 lb) of faecal waste, and require a further 1.5 kg (3.3 lb) of water for washing purposes. The logistic implications of storing and processing this weight and bulk on board a spacecraft are quite apparent and yet they have been accepted for the relatively short-duration missions so far undertaken. This is because the management systems involved in the present generation of life support systems are straightforward and reliable if not simple, and the orbiting space laboratories manned so far have all been amenable to replenishment by craft sent up from Earth. And the Apollo missions to the Moon were of sufficiently short duration that all consumables could be adequately provided in the form of expendable supplies loaded on Earth. When extended missions beyond the ability of re-supply from Earth are undertaken, however, the weight and bulk penalties of stored supplies will be intolerable. Long-term missions will therefore have to employ methods of regenerating the vital elements of a life support system (LSS).

The perfect solution would be to have a completely closed LSS in which all matter was conserved, processed, and re-cycled. But this is clearly not possible since spacecraft are never completely sealed and some nutritional needs cannot be met by closed systems. It is possible, however, to regenerate or reclaim water and the principal atmospheric gases (although the oxygen consumed by man must be replaced), and to dispose of waste products. Furthermore, such a scheme can be integrated so that the by-products of one sub-system provide the substrate for the needs of another: a so-called semi-closed integrated regenerative life support system. Proven technology could

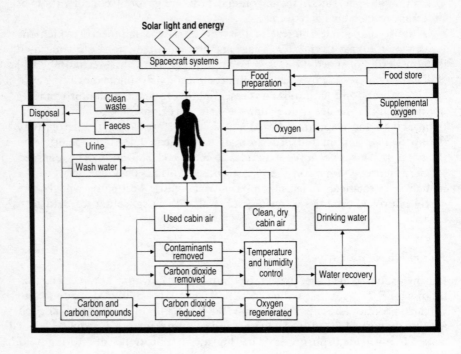

Figure 13.1 A simple physico-chemical regenerative life support system

already support a system to do this, based on physico-chemical processes, and Figure 13.1 shows a simplified arrangement of a semi-closed regenerative LSS.

Atmospheric control   As elevated levels of carbon dioxide in the inspired air are physiologically unacceptable, that gas must be removed from the atmosphere of a closed environment for safety reasons. But oxygen can be regenerated from carbon dioxide, and so the subsequent manipulation of the latter holds the key to atmospheric revitalization and hence to the success of a semi-closed LSS. Several methods are available which will scrub the atmosphere of carbon dioxide and then allow it to be used in other processes. One such is the molecular sieve which, under the cyclic influence of pressure, first adsorbs the gas and then desorbs it. Water is preferentially adsorbed by the zeolite material of the sieve and so two units working in parallel are usually employed: the first to remove water and the second carbon dioxide.

Molecular sieves require heat and vacuum sources, and do not perform very well when the pressure of carbon dioxide (PCO_2) falls below about 5 mmHg (0.7 kPa). Nevertheless, as described in Chapter 3, they were successfully used on Skylab and the Russians have demonstrated their effectiveness in very long, Earth-based studies. Amine scrubbing systems are also potential candidates for carbon dioxide collection and have the advantage of working at lower temperatures and of coping with low values of PCO_2. An even better method would be the so-called hydrogen depolarized cell, a continuous low temperature electrochemical process with no moving parts. This will also cope with low PCO_2 levels, is more economical in terms of weight and, since the process involves a fuel-cell type of reaction between oxygen and hydrogen which produces water *and* some electricity, no external power source is needed!

All of these processes are able to pass on carbon dioxide gas to the next stage in the cycle: catalytic reduction with hydrogen to produce water. Once again, several methods are available for exploitation. Both the Sabatier and Bosch processes, whereby carbon dioxide is heated with hydrogen gas in the presence of catalysts (usually metals) to produce water and methane, and water and solid carbon respectively, are established technologies which could be adapted for use in space. The former has an additional advantage in that it is an exothermic (heat-producing) reaction and the heat can be used to burn off toxic contaminants including the unwanted methane.

The final step in the revitalization process is the electrolysis of water to produce oxygen gas. Many electrochemical methods of extracting oxygen from water exist, including the use of salts or alkalis in aqueous solution, the use of ion exchange membranes, and the use of solid electrolytes such as polymers. The last would be particularly attractive since it is non-caustic, has a high output for a low energy requirement, and is very efficient. Other solid electrolytes, however, may provide an even greater benefit if they can be made to work by directly decomposing carbon dioxide into carbon and oxygen, so eliminating one stage from the cycle: the use of fused carbonate salts for this purpose has met with some success. In this process, which is extremely difficult to engineer, cabin air passes directly into the cell which then absorbs and dissociates carbon dioxide and water, deposits pure carbon at the cathode and produces oxygen. The oxidizing power of the unit should even cope with gaseous cabin contaminants. Once regenerated, the oxygen must be supplemented with virgin gas, to make good the metabolic consumption, and then mixed with the inert component (the level of which remains constant since it is not consumed) before passing back into the cabin. Nitrogen is physiologically appropriate and would be the most obvious choice, although the use of other inert gases such as argon or helium would reduce the risk of decompression sickness (chapter 3). Alternatively, as in the early generations of American spacecraft, a reduced cabin pressure could be employed to minimize this risk. Other benefits outweigh this single indication, however, so that the Russian space station Mir enjoys an Earth-life atmosphere in terms of composition and pressure, and the American Space Station will do likewise, as indeed does the Shuttle.

Water Man's requirement of water represents, by weight, nearly half of his life support needs, and it is clear that the efficient management and reclamation of water would be of enormous logistic benefit. It will be recalled that the fuel cells used for energy production on board many spacecraft, including the Shuttle, produce large quantities of water as a by-product (Chapter 7). Furthermore, the quality of such water is high and very little additional processing is needed to render it potable. In the future, however, there may be a need to recycle the water used for washing purposes and even that produced as urine and in faeces. As with atmospheric recycling, several options for water management are available to the space engineers. For example, wash water can be effectively cleansed by reverse osmosis: a technique whereby dirty water is first passed through particulate filters capable of trapping bacteria and then forced through selected semi-permeable membranes which are able to remove virtually any dissolved material. Such pressure filtration requires little energy and can cope with large quantities of fluid. The chosen method of reclaiming pure water from urine, however, is vacuum distillation followed by pyrolysis (chemical decomposition by heat) and then condensation. Pyrolysis of the distillate vapour ensures that organic contaminants and bacteria are destroyed. But other methods and combinations of methods exist, and the final choice will depend upon such factors as system energy and heat requirements.

The extraction of water from faeces not only supplements the recycled fluid volume but also reduces the weight, the bulk, and the microbial activity of solid waste material. Once in the dehydrated form, solid waste can be sterilized before being discharged overboard or stored for later disposal. Eventually, it may also be possible to extract useful solid components for re-use. Once reclaimed, the water from whatever source must be suitably treated prior to storage and later use either for consumption or ablution. Heat clearly achieves a considerable level of sterility, but chemical additives, such as iodine or silver salts, are also necessary although complete sterility can probably never be achieved in space. And of course, palatability and other aesthetic aspects (such as colour and odour) must be remembered; this is especially so if recycled water is to be used as a vehicle for dietary supplements.

Thermal control and humidity Another aspect of the semi-closed LSS will be the provision of adequate thermal control and humidity. The accomplishment of these needs is of more concern to the spacecraft engineer than to the physiologist who merely states what is required of the engineering in order to maintain the occupants in comfort.

Food The savings in weight achieved by atmospheric revitalization and water reclamation will be of major and necessary benefit to future space missions. The potential to save further weight by on-board production of

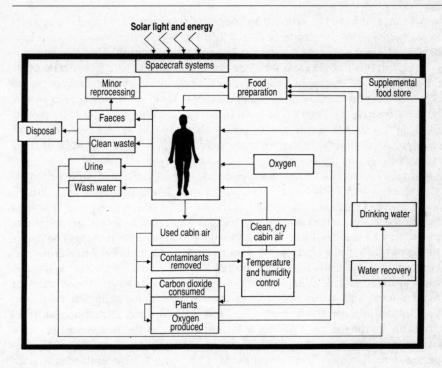

Figure 13.2 Integration of a biological element within a simple regenerative life support system

food is also under active investigation but is unlikely to be pursued too vigorously in the foreseeable future because of the immensely powerful role food plays in the physical and psychological well-being of man. The weight and bulk of food can be and is optimized on Earth by careful preparation and packaging for space (Chapter 7), and the evolution of food technology is likely to remain largely terrestrial. If some food production could be achieved in flight, however, then the LSS would become even more autonomous, although perhaps also more susceptible to the external influences of space and consequently less reliable. The ever-present problems of plant disease and genetic mutation, either of which could devastate a crop, would add to the vulnerability of such a system. Figure 13.2 shows how a biological element could be integrated with the physico-chemical system discussed above, so creating a Closed Ecological Life Support System (CELSS).

There is no obvious reason why plants from Earth should not be grown successfully during spaceflight but, as with human function, the effects of microgravity on plant physiology have yet to be fully established. The elegantly simple botanical studies of plant germination and growth carried out on Skylab, Salyut, and Spacelab were, therefore, of the utmost importance to the future of man's long-term occupation of space and such experiments must continue if closed life support systems are to develop. Many different types of foodstuff are potentially suitable for use in such systems but it is not yet clear whether plants can survive and flourish through multiple growth cycles in space. Nevertheless, since photosynthetic plants consume carbon dioxide and produce oxygen, their use within a LSS would be an attractive means of 'completing the (atmospheric control) loop' and forming a CELSS. Such plants could also provide at least some of the required food supply, thereby greatly reducing the weight and bulk of food stored conventionally.

Current research in this area has concentrated on rapidly growing primitive plant forms, such as the algae Chlorella sp and Chlamydomonas sp, although some higher forms, including Spirodela polyrhiza (duckweed) and Ipomoea batatas (sweet potatoes), have been studied. Algae can be grown in fermenters and it may be that the biotechnology required in space for these will be easier to accomplish than that for the more appetizing conventional crop and vegetable plants, which require pollination, are slow growing, and need cyclical illumination. With regard to higher plants, however, the sciences of hydroponics and aeroponics (plant growth within nutrient-rich fluid and vapour media respectively: that is, soil-less) offer distinct attractions, and will probably be the methods of choice for agriculture in the large space colonies of the next century. Finally, the non-photosynthetic bacterium Hydrogenomonas has also been investigated with considerable enthusiasm since cultures of this micro-organism can convert the hydrogen, carbon dioxide, and urea from human waste into potentially useful food products.

Artificial gravity

Current evidence suggests that the physiological changes associated with exposure to microgravity will give rise to serious consequences for the returning space traveller if the flight duration exceeds eight or nine months. The crew members of Soyuz T10 took several months to recover overtly from their 237-day sojourn aboard Salyut 7 despite the rigorous countermeasures adopted; and of particular concern was the persistence of neuro-muscular dysfunction. Cardiovascular and neuro-vestibular deconditioning were also marked, and the possible long-term effects on the musculo-skeletal system remain of great significance (see Chapter 11). The proposed length of continuous stay on board Space Station is 90 to 120 days and, although repeated tours may alternate with suitable recovery periods on Earth, such exposure will probably be safe.

But a manned mission to our nearest planetary neighbour, Venus, could take a total of 800 days; while the more likely first interplanetary journey, to Mars, could have a total mission length of over 1,000 days. Even though both of these estimates would include a period of about 450 days on the planet's

Table 13.1 Physical characteristics of some planetary environments

Planet/ satellite	Atmosphere Pressure (atmosphere)[1]	Composition	Surface Temp (°C)	Gravity	Period of Rotation (Earth Days)	Distance from Earth (million km)	Approximate Journey Time (Earth Days)
• Earth[2]	1	21%O_2, 79%N_2	+50 to −88	1 G	1	–	–
• Moon	0	–	+110 to −178	0.17 G	27.3	0.4	3
• Venus[2]	90	95%CO_2, 5%N_2	+450	0.91 G	243.0	39 – 260	175
• Mercury[2]	0.001		+340	0.38 G	58.6	80 – 220	170
• Mars[2]	0.005	>90%CO_2	+20 to −70	0.38 G	1.02	56 – 400	270
• Jupiter	200,000		−140	2.6 G	0.41	588 – 963	720

Notes: 1 One atmosphere = 760 mm Hg (101.3 kPa)
  2 Comprise the terrestrial planets, composed of rock and iron

surface, the journey times are approaching the limit of physiological acceptability for exposure to microgravity. Furthermore, although Venus has a gravity of 0.9G, its atmosphere is composed predominantly of carbon dioxide at a surface atmospheric pressure of ninety times that of Earth! The surface pressure on Mars is a more acceptable 0.005 of that on Earth, but the atmosphere is again composed primarily of carbon dioxide, and gravity is only 0.38G. Table 13.1 summarizes the leading environmental features of some major planetary bodies, and demonstrates the enormity of the task confronting the explorers of the future: the requirement for both CELSSs and artificial gravity is self-evident.

Artificial gravity will thus be a possible requirement for long-term occupation of an Earth orbital craft and, if the voyagers are ever to return to Earth, an essential requirement for interplanetary missions. For the same reason, it may be that colonies established on the surface of some planets may also require facilities for increased gravity. At present, the only practical method of inducing a suitable gravitational field on board a spacecraft is to rotate the entire vehicle. On-board man-carrying centrifuges are a possible alternative, but constraints of size would dictate a short radius of rotation and therefore relatively high rates of turn, so perhaps creating more problems than solved. (Small on-board centrifuges do have a place in life sciences research, however, since they provide control conditions for the study of animal and plant behaviour in space. Such centrifuges will form part of the hardware provided for scientists on board space stations, although their operation will naturally affect other experiments requiring stable and profound levels of microgravity: hence the powerful arguments propounded by astronomers and material scientists in support of unmanned space platforms.) Rotation of a spacecraft in flight, even if intermittent, will obviously have repercussions on its overall design, its propulsive mechanisms, and its navigation. And, although intended to minimize physiological deconditioning, it may also give rise to other problems including, most importantly, the effects of varying gravity levels on neuro-vestibular function, and the transition to and from the different levels. Space motion sickness may be exacerbated, perceptual disturbances may be profound as it becomes easier to walk in one direction than another, and there will be a tendency to fall in one direction while ascending and the opposite direction while descending. The optimal levels of such rotation will require much research, but a level of 0.3–0.5G, in combination with other countermeasures (Chapter 11), has been suggested as sufficient to prevent major cardiovascular and musculo-skeletal deterioration.

For space colonies established as free-flying structures above other planets – for it may be prudent not to live on the surface of other worlds – many of the constraints of size, engineering, and mobility are removed (Plate 15). Vast rotating structures could be built which would generate an Earth-like gravitational field at the outer edges, while those activities requiring low levels of gravity would be undertaken near to or at the axis of rotation. Radial acceleration of a body is given by the equation a = v^2/r (where a = acceleration, v = linear velocity of a surface point, and r = radius). The required acceleration for 1G is 32 ft.sec^{-1} (9.8 m.sec^{-1}) (chapter 4) and so a realistic, cylindrical space colony with a radius of 328 ft (100m) would need to rotate at a velocity of 102.4 ft.sec^{-1} (31.2 m.sec^{-1}) or 6,144 ft.min^{-1} (1,873 m.min^{-1}, (70 mph)). Since a point on the surface of such a construction would move through $2\pi r$ or 2,060 ft (638 m) in one revolution, the colony would be spinning at three revolutions per minute. (An unacceptable spin rate of nearly nineteen per minute would have been necessary to create this degree of artificial gravity on board Skylab, with its radius of 11 ft (3.3 m); hence the suggested reduction in G level for small craft in transit.) In order to counteract the displacement effect caused by spinning, two such cylinders rotating in opposite directions would be needed. Living and recreation areas, as well as the farming support required, could all be carried out within the cylinders; while supporting workshops and other facilities could be sited in

Table 13.2 Some clinical hazards of spaceflight

Illness as a consequence of being in space:
- Space adaptation syndrome
- Hypoxia
- Oxygen toxicity
- Decompression sickness
- Radiation sickness

Illness/Injury as a consequence of working in space:
- Electric shock: burns
        cardiac dysrhythmias
- Trauma: minor (lacerations, contusions)
        : moderate (deep cuts, foreign
            bodies, concussion)
        : major (fractures of long bones,
            ribs, skull and spine;
            penetrating injuries to
            viscera)
- Thermal injury: heat exhaustion
        : frostbite
- Burns: major, minor, and ultraviolet
- Toxic contamination: including dermal,
        neurological, gastrointestinal
        and pulmonary effects

Illness incidental to the space environment:

Clearly, all terrestrial diseases which afflict humans could manifest in space, although those with pre-existing conditions, such as diabetes mellitus or epilepsy, will be excluded from spaceflight for the immediate future. Sudden cardiovascular, cerebro-vascular and abdominal emergencies (myocardial infarction (heart attack), cerebral thrombo-embolism (stroke), and appendicitis respectively, for example) are obviously of particular importance; as are local or general infections, and dental problems.

external pods. If each cylinder was 0.62 mile (1 km) long, a population of just over 10,000 could be supported!

In-flight health care

Unless extensive medical facilities are available on board spacecraft designed for long-term occupation, either in near-Earth orbit or on deep space exploratory missions, serious injury or illness in space may well prove fatal long before a rescue mission could be mounted. Table 13.2 lists some of the potential clinical problems which must be considered.

Health care in space is, therefore, of paramount importance and the facilities it requires must be equivalent to those provided by a hospital dealing with emergencies on Earth. The standard medical practices of prevention, diagnosis, and treatment must all be supported by appropriate equipment and personnel skilled in their use. Thus, normal standards of terrestrial environmental health must be adopted wherever possible so that risk of injury or illness from surrounding hazards is minimized. Electrical and mechanical equipment must be designed and operated with safety the principal consideration; sanitary arrangements must likewise be carefully engineered and managed; and the overwhelming advantages of prevention rather than cure must be emphasized to the crew. Laboratory facilities must be provided to allow routine monitoring of human health (clinical biochemistry, haematology, immunology, microbiology, toxicology, and urinalysis) and that of the life support systems (organic and inorganic contamination). Clinical diagnostic equipment must also be available for both physical examination (including stethoscope, sphygmomanometer, ophthalmoscope, and auroscope) and supplementary investigation including telemetry of data to Earth, such as ECG, EEG and diagnostic imaging (X-ray, computerized tomography and ultrasound). Treatment of minor injury and illness will require an adequate pharmacy stocked with appropriate medical and dental instruments, dressings, and medications. More serious conditions may require surgery and so a fully equipped operating theatre will also be needed, while the ability to perform effective cardio-pulmonary resuscitation is another obvious need. A hyperbaric chamber, capable of recompressing a victim of decompression sickness, would complete the essential components of a health care facility in space. A secondary role of such a facility would be to support any on-board medical research programmes.

Personnel trained both in primary health care and in research methods will thus be essential members of the crew of any large spacecraft on a long-term mission. For such missions it is likely that at least one crewman will be a physician trained in surgery and dentistry, and that a second doctor, perhaps an anaesthetist trained in intensive care, will be carried as well. The space colonies of the future will naturally require even greater hospital facilities, including those for gynaecology, obstetrics, paediatrics, and psychiatry.

Some final thoughts

Although strictly outside the remit of this book, it is worth remembering that considerable benefits have accrued to mankind in general as a consequence of the manned spaceflight programmes; and that the implications of these so-called spin-offs have, and will continue to have, an enormous influence which is often not immediately apparent. The technology required for spacecraft systems has always yielded benefits for terrestrial science. Computer evolution is a prime example of this, and the ubiquitous pocket calculator, as well as the increasing sophistication of micro-computers for home use and myriad software programmes, are directly attributable to the space programme. Similarly, the need to communicate continuously and reliably with manned spacecraft led to the development of a network of communications satellites from which the entire world now benefits. Many other, perhaps more mundane, examples exist: smoke detectors for the home, insulating materials for liquid gas tankers, tin can crushers for use when camping, water purifying filters, heat shields for furnaces and aircraft, and the oft-quoted invention of Teflon. For the art of medicine, the list of spin-offs is equally impressive and includes: the miniaturization of physiological monitoring and recording equipment, the development of non-invasive techniques of assessment (a digital blood pressure monitor was flown on a recent Spacelab mission), a low intensity X-ray imaging system which reduces exposure to X-rays by over 99 per cent, an implantable cardiac defibrillator and programmable pacemaker, a tissue stimulator which can be used to control pain and neuromuscular disorders, an eye movement recorder and automatic analyser, a non-invasive bone strength analyser which is being refined for diagnostic use in bone diseases, fibre optic technology and reading machines for the blind, helmet-mounted cameras for teaching surgery, robot arms and voice-controlled wheelchairs for use by tetraplegics, and lightweight thermal blankets. In addition, microgravity offers the clinically unique opportunity to study induced changes in physiology in the absence of disease. The opportunity to carry out such studies will lead to a better understanding of disease processes on Earth, and so to a greater likelihood of successful palliation or cure (Plate 14). The loss of bone mineral in microgravity, which so closely resembles the condition of osteoporosis, is the most obvious example of this. While it is true that most of these benefits could have been developed eventually, it is extremely unlikely that so many of them would have been so vigorously pursued in the absence of the driving needs of the manned spaceflight programmes. Furthermore, medical and industrial research both on Earth and in space will continue to provide great advances in knowledge for the good of all.

Finally, of course, space offers a classroom literally as big as the Universe; and it is to be hoped that future generations of schoolchildren will enjoy learning about the mysteries of their Earth and the cosmos during lessons transmitted from space. In so doing they will become aware of space and of its vast potential: characteristics which I believe to be essential for our future survival on Earth as well as in space.

Bibliography

Many books and original papers were consulted during the preparation of this book. The following list includes some of the former which I found to be of particular interest and importance.

Baker, D. (1981) *The History of Manned Spaceflight,* London: New Cavendish Books.

Calvin, M. and Gazenko, O. G. (eds) (1975) *Foundations of Space Biology and Medicine: I Space as a Habitat: II Ecological and Physiological Bases of Space Biology and Medicine (two books): III Space Medicine and Biotechnology*, joint USA/USSR publication, Washington, DC: NASA SP-374.

DeHart, R. L. (ed.) (1985) *Fundamentals of Aerospace Medicine*, Philadelphia: Lea & Febiger.

Despopoulos, A. and Silbernagel, S. (1981) *Colour Atlas of Physiology*, Stuttgart: Georg Thieme Verlag.

Engle, E. and Lott, A. (1979) *Man in Flight: Biomedical Achievements in Aerospace*, Maryland: Leeward Publications.

Ernsting, J. and King, P. F. (eds) (1988) *Aviation Medicine,* 2nd edn, London: Butterworths.

Furniss, T. (1983) *Manned Spaceflight Log*, London: Jane's Publishing Company Ltd.

Harding, R. M. and Mills, F. J. (1988) *Aviation Medicine,* 2nd edn, London: British Medical Journal.

Joels, K. M., Kennedy, G. P., and Larkin, D. (1983) *The Space Shuttle Operator's Manual*, London: Papermac.

Johnston, R. S. and Dietlein, L. F. (eds) (1977) *Biomedical Results from Skylab*, Washington, DC: NASA SP-377.

Johnston, R. S., Dietlein, L. F., and Berry, C. A. (eds) (1975) *Biomedical Results of Apollo*, Washington, DC: NASA SP-378.

Lewis, R. S. (1983) *The Illustrated Encyclopedia of Space Exploration*, London: Salamander Books Ltd.

Mason, J. A. and Johnson, P. C. (eds) (1984) *Space Station Medical Sciences Concepts*, NASA Technical Memorandum 58255.

NASA (1983) *The First 25 Years: 1958–1983: a resource for teachers*, Washington, DC: NASA EP-182.

Nicogossian, A. E. and Parker, J. F. (1982) *Space Physiology and Medicine*, Washington, DC: NASA SP-477.

Novosti Press Agency (1986) *Road to the Stars*.

Pogue, W. R. (1985) *How Do You Go to the Bathroom in Space?*, New York: Thomas Docherty Associates.

Turnill, R. (ed.) (1986) *Jane's Spaceflight Directory*, London: Jane's Publishing Company Ltd.

Wilson, A. (1982) *The Eagle has Wings: the story of American space exploration 1945–1975*, London: The British Interplanetary Society.

Wolfe, T. (1980) *The Right Stuff*, London: Jonathan Cape Ltd.

(NASA publications are for sale by the Superintendent of Documents, Government Printing Office, Washington, DC 20402.)

In addition, the following journals regularly publish original articles, reviews, and items of general interest in the field of manned spaceflight:

Aviation Space and Environmental Medicine, the journal of the Aerospace Medical Association (monthly).

Journal of the British Interplanetary Society (monthly).

Spaceflight, another publication of the BIS (monthly).

Flight International (weekly).

Aviation Week and Space Technology (weekly).

Glossary of some medical terms

Aerodontalgia	Pain in a tooth as a consequence of a reduction in barometric pressure.
Aetiology	The origin or cause of a medical condition.
Anorexia	Loss of appetite.
Antibodies	Defensive protein materials developed by the body, usually in response to the presence of foreign substances such as bacteria.
Ataxia	Lack of co-ordination of muscular action.
Atrophy	Reduction in the size of an organ or cell which is, or has been, in disuse.
Cataract	Opacity of the lens of the eye.
Cohort study	The study of a group of persons over a prolonged period of time.
Congenital	Originating before birth.
Conjunctiva	The thin mucous membrane which lines the eyelids and is reflected to cover the front of the eyeball.
Cyanosis	Blue discoloration of the skin due to the presence of abnormal amounts of de-oxygenated haemoglobin in the blood.
Defaecation	The act of opening the bowels to evacuate faeces.
Degeneration	The deterioration in structure, or impairment of function, of a tissue or organ.
Deglutition	The act of swallowing.
Demineralization	The process whereby salts are lost from the body in excessive amounts.

Depersonalization	The process by which an individual suffers a morbid loss of sense of identity.
Dermatitis	Inflammation of the skin, particularly when associated with redness, itching, and flaking.
Diuresis	Increased excretion of urine.
Dysfunction	An absence of completely normal function.
Dysrhythmia	A disturbance in the normal cardiac rhythm.
Electrocardiogram (ECG)	A graphic record of electrical activity generated by the heart
Electroencephalogram (EEG)	A graphic record of electrical activity generated by the brain.
Electromyogram (EMG)	A graphic record of activity generated by contraction of skeletal muscle in response to an electrical stimulus.
Electro-oculogram (EOG)	A graphic record of the electrical activity generated by the muscles which control movements of the eye.
Endocrine	Pertaining to the system of glands which produce hormones.
Extensor	A muscle that extends a part of the body.
Flexor	A muscle that bends a part of the body.
Gangrene	Death and putrefaction of a soft tissue.
Haemoconcentration	A relative increase in the number of circulating red blood cells as a consequence of a reduction in the volume of plasma.
Haemoglobin	The iron-containing, oxygen-carrying pigmented protein molecule contained within red blood cells (and responsible for their colour).
Haemopoeisis	The process by which blood is formed.
Haemostasis	The arrest of bleeding (or of circulation).
Halitosis	Offensive breath.
Homeostasis	The process by which the body's internal environment is maintained in a state of equilibrium.
Homogeneous	Being of uniform composition or structure.
Hydrostatic pressure	The pressure exerted by a liquid when in equilibrium and at rest.
Hyperarousal	A state of heightened mental and physical awareness.

Hyperventilation	An increase in the rate and/or depth of respiration such as to exceed the body's requirement to eliminate carbon dioxide.
Hypothermia	A condition in which the body temperature is lower than normal.
Hypoxia	A deficiency of oxygen.
Intracranial	Within the cranium or skull.
Lachrymation	The secretion and flow of tears.
Macerate	The process of softening a tissue by soaking in a fluid.
Mastication	The action of chewing food.
Matrix	1. The substance found between the cells of a tissue.
	2. The place in which development occurs.
Oedema	A condition in which body tissues contain an excessive quantity of fluid, and are consequently swollen.
Olfaction	The sense of smell.
Orthostatic intolerance	An inability to tolerate the upright posture, with a consequent tendency to faint.
Osteoporosis	A condition in which the bones become increasingly porous and brittle as a result of loss of mineral constituents. It is a normal accompaniment of the ageing process.
Palliation	The process of relieving or alleviating a condition without effecting a cure.
Paraesthesiae	Abnormal sensations of numbness, prickling and tingling.
Percentile	One of ninety-nine values of a variable which divide a population into 100 equal groups with regard to the value of that variable.
Perfusion	The process by which a fluid (e.g. blood) passes through an organ or tissue.
Physiology	The scientific study of the normal functions of cells, tissues, organs, and systems of living organisms.
Pneumonitis	Inflammation of the lung.
Potable	Suitable for drinking.
Prophylaxis	Application of treatment or procedures to prevent disease.

Proprioception	The awareness of the body's posture, movement and changes in equilibrium, and the knowledge of position, weight, and resistance of objects in relation to the body. Achieved by means of specialized sensors (proprioceptors) in the joints, muscles, and tendons as well as by the organs of special sense (vision and balance).
Rubella	German Measles: an acute viral infection.
Serology	The scientific study of serum reactions, diagnosis and treatment.
Spasticity	A condition of abnormally increased muscular tension causing stiff and awkward movements.
Tracheostomy	The operation of cutting into the trachea (windpipe) so allowing the insertion of a tube to overcome tracheal obstruction.
Transudation	The process whereby fluid passes through the pores or interstices (gaps) in a membrane.
Vasoconstriction	Constriction (narrowing) of blood vessels.
Vasodilatation	Dilatation (widening) of blood vessels.
Viscera	The internal organs, particularly those of the abdomen (but also of the skull and chest).

Name index

Subject index

acceleration 60–1
  profiles 67
accelerations
  linear 62
  long duration 62
    physiology of +Gx 66–8;
    physiology of –Gx 68;
    physiology of +Gz 62–4, 191;
    physiology of –Gz 64–5
  radial 62
  short duration 65–6
aerodontalgia 46
aeroponics 202
Agena spacecraft 21
aids to mobility 120–3
Air-Launched Sortie Vehicles
  (ALSV) 18, 34
aldosterone 162–3
algae 202
alkaline superoxides 49, 51
altitude, ascent to,
  physiological effects of 48;
  pressure changes on 42–7;
  relation with oxygen partial
  pressure 43
American manned spaceflight
  programme 18–35
amine scrubber 50, 199
amino acids 93
angiotensin 162
Anthrorack 32
antibiotics, resistance to 167
antibodies 27, 166
antidiuretic hormone (ADH) 162
anti-emetic drugs 155
anti-g suit 64, 181
Apollo Applications Programme
  see Project Skylab

Apollo Extravehicular Mobility
  Unit 56, 59
Apollo Launch Escape System 24,
  70
Apollo lunar landing programme
  24–28
Apollo Lunar Surface Experiment
  Package 22, 26–8
Apollo Portable Life Support
  System 25, 27, 53, 56–7, 87
Apollo spacecraft 22–4;
  acceleration profile 8, 67;
  carbon dioxide levels 53;
  configuration 22–4, 29;
  emergency escape 24, 70;
  food supply 95–6;
  humidity control 91;
  landing system 69;
  life support system 25, 52–3
    portable 25, 27, 87, 116;
  power supply 102;
  radiation protection 78;
  thermal control 85;
  waste-management system 104–6;
  water supply 95, 102
Apollo 1 25, 52, 115
Apollo 7 25, 171
Apollo 8 25, 171, 128
Apollo 9 25, 128
Apollo 10 25–6
Apollo 11 26, 54, 73, 135
Apollo 12 26, 69, 73, 102, 104, 134,
  135
Apollo 13 27, 53, 59, 73, 85, 128,
  134
Apollo 14 27, 105, 130, 131, 135
Apollo 15 27–8, 102, 131, 133, 183
Apollo 16 28, 56, 102, 105

616.9
Har

Harding, Richard

Survival in space

S1976

88-32091

$27.50

616.9
Har

Harding, Richard

Survival in
space

S1976

88-32091

$27.50

DATE	BORROWER'S NAME	